In His
IMAGE

In His IMAGE

Finding Courage, Hope, and Confidence to Become More Christlike

STEVEN A. CRAMER

Covenant Communications, Inc.

Published by Covenant Communications, Inc.
American Fork, Utah

Printed in the United States of America
First Printing: August 1999

06 05 04 03 02 01 00 99   10 9 8 7 6 5 4 3 2 1

ISBN 1-57734-583-5

For the body is not one member, but many.

If the foot shall say, Because I am not the hand, I am not of the body;
is it therefore not of the body?

And if the ear shall say, Because I am not the eye, I am not of the body;
is it therefore not of the body?

If the whole body were an eye, where were the hearing?
If the whole were hearing, where were the smelling?

But now hath God set the members every one of them in the body,
as it hath pleased him.

And if they were all one member, where were the body?

But now are they many members, yet but one body.

And the eye cannot say unto the hand, I have no need of thee:
nor again the head to the feet, I have no need of you.

Nay, much more those members of the body,
which seem to be more feeble, are necessary:
And those members of the body, which we think to be less honourable,
upon these we bestow more abundant honour;
and our uncomely parts have more abundant comeliness.

For our comely parts have no need: but God hath tempered the body together,
having given more abundant honour to that part which lacked:

That there should be no schism in the body;
but that the members should have the same care one for another.

And whether one member suffer, all the members suffer with it;
or one member be honoured, all the members rejoice with it.
(1 Corinthians 12:14-26)

TABLE OF CONTENTS

INTRODUCTION

One rainy day I saw a mother and a young boy come out of a store. As the mother hurried to her car, obviously preoccupied with thoughts of her next task, the little boy stayed behind and began stomping in a puddle. I watched as he repeated this enjoyable activity over and over, trying each time to make a bigger splash. Soon, however, his fun was interrupted. The mother, discovering that the boy was not with her, yelled at him to "get in right now."

Unfortunately, many well-meaning disciples of Christ are like the carefree boy. They choose their priorities and live their daily lives as though this important earth-school were a playground for our pleasure and entertainment instead of a spiritual battleground for the souls of men.

We come to *this* world to prepare ourselves for the *next* world. Thus the Lord has stated that "if you will that I give unto you a place in the celestial world, *you must prepare yourselves* by doing the things which I have commanded you and required of you" (D&C 78:7). Every ten seconds the world's population increases by 27 people. (See *National Geographic Magazine*, Oct. 1998, editorial page.) That means that well over 200,000 people a day join us in the spiritual challenge of finding our way back to Heavenly Father. We come here not just to *obtain* a physical body, but also to learn how to *use* it wisely. We must use it to produce joy and increase our capacities instead of succumbing to the captivity and restrictions that our fallen, natural-man flesh tries to impose.

We all live in physical bodies, but we don't all have the same physical experiences. Some of us consider ourselves ugly because of weight, size, or appearance problems. Some of us wish our bodies were taller or shorter, thinner or heavier, more muscular or more dainty. We wish our hair were longer or shorter or a different color. (Or that we just had some.) Some of our bodies are old and weak. Some have high cholesterol, high blood pressure, diabetes, or other limiting factors. Some of us are trapped in bodies that are blind, deaf, crippled, deformed, or paralyzed. Instead of appreciating the lives and bodies Heavenly Father gave us, many of us look with envy at the tabernacles that seem more desirable, violating one of the fundamental laws he gave for spiritual progress: "Thou shalt not covet . . . any thing that is thy neighbour's" (Ex. 20:17).

The miserable captivity suffered by a small girl named Cindy Erickson represents our mortal dissatisfactions as well as any I've heard. At the age of seven months, Cindy's tiny body was attacked by a rare genetic disease called glutaric acidemia. This disease racked her body with a ceaseless twisting and contraction of her muscles. This constant movement rendered her not only helpless, but almost unable to receive help from others. Sometimes her mother spent seven hours a day just trying to feed her. She couldn't walk or talk her entire life. Because of the way her body tortured her, she cried day and night and was rarely able to sleep for more than 45 minutes in a 24-hour period. She died at the age of twenty, her pitiful body weighing only 50 pounds. (See Bruce and Joyce Erickson, *When Life Doesn't Seem Fair* [Salt Lake City: Bookcraft, 1995, pp. 280-81] or *Ensign*, Jan. 1999, p. 10.)

We encounter endless physical defects and difficulties in our mortal, imperfect bodies. But through it all, the supreme issue is not our *appearance* or physical limitations, but what we do with our minds, hearts, and spirits. Do we allow physical impairments to embitter us and diminish our feelings of self-worth? Or do we find and apply gospel truths so that our spirits develop toward godliness and happiness?

You and I fought a war in heaven for the right to come here and choose what to do with our bodies during this mortal schooling. Trying to think of a sin you can commit or a good deed you can do without using your mind or your body makes it clear that everything that happens to us in this life is affected by the way we use our bodies—or the way we allow our bodies to use us.

Each choice we make, each activity we perform with the various parts of our *physical* anatomy, affects our *spiritual* anatomy, and therefore, our spirituality and spiritual destiny. Because of the choices we make, some of us live in temple-bodies, where the Spirit of the Lord dwells in rich abundance. On the other hand, some of us succumb to the appetites and desires of the body, causing us to struggle through mortality in prisons of flesh that are enslaved by the desires, passions, and habits that rule our lives. Each choice we make for the way we use our time and our bodies leads us *upward* toward greater spirituality, freedom, and self-mastery, or *downward* toward enslavement and sorrow. The choices we make for our bodies will determine whether our mortal probation sends us to an eternal reward of joy in the celestial kingdom or to a wasted eternity in a lower kingdom.

In the beginning, Genesis tells us, God created us in his image *physically*. But the creation is not over. Our challenge during mortality is to partner with God to create ourselves in his image behaviorally, morally, mentally, emotionally, spiritually, and every other way. This creation of self means overcoming the selfish, carnal orientation of our fallen bodies. It means we must rise above the animal plane of carnal desires to the spiritual values that God holds dear. It means mastering our physical bodies so that we become truly holy, godly, and Christlike. It means, in his words, to "follow me, and do the things which ye have seen me do" (2 Ne. 31:12; see also John 13:15).

This challenge to master our bodies of flesh, to have dominion over every unworthy desire, and to control bodily behavior so that we never offend the Lord's Spirit can seem overwhelming. It is an enormous challenge when viewed as a *whole*. However, when we break it down to training and controlling the action of *each individual member* of the body, such as our eyes, ears, mouth, and so forth, it is much easier to accomplish.

The scriptures contain thousands of verses revealing the dangers our spirits face while living within bodies of fallen flesh. They also present special strategies and instructions for using each member of the body in ways that will unite the flesh and the spirit person inside in a common pursuit of sanctification and exaltation. If we ignore these instructions for our spiritual anatomy, we will find ourselves prisoners to the carnal flesh, spiraling downward toward captivity to the lusts of the flesh and forfeiture of our rightful place in the next life. If we do our best to apply them, we make it possible for Christ to rescue us from the fall and give us a newness of life in his image.

The purpose of mortal probation is not to change *from* who we are, but to change back *to* who we *already* are inside: the divine off-spring of God, eternal spirits involved in a schooling process that can lead to eternal glory and joys beyond our present comprehension. If your life is not as spiritual and victorious as you wish it were, don't be afraid to look at where you are in your present defeats. Admitting where you are spiritually right now is not a failure, but a doorway to improvement. The path to achieving this improvement is to simply do with each part of your body what you think Jesus would do in the same circumstance.

In this book, we'll learn how to make temples of our bodies. We'll learn how to use each part of our bodies as instruments of service, joy, and fulfilment so that we can return to our Heavenly Father with honor and with the image of his Only Begotten Son.

(Note: All italics throughout the book are the author's added emphasis unless otherwise noted in the citation.)

Chapter One

THE BODY

One of my seven daughters is blind from diabetes. Her seeing-eye dog is named Eddy. I never thought of Eddy as a "natural-man" dog until one day when Jeri and I were walking along a sidewalk and I saw a lady approaching with another dog on a leash. I knew exactly what was going to happen, because it's just a part of the nature of dogs to sniff and check each other out. I don't care what orders you give or how hard you pull on the leash; you probably won't be able to move on until the dogs are through getting acquainted.

Eddy has some other "natural-dog" traits you'll recognize. If someone comes to visit, it's almost impossible to do so until Eddy has received his share of the attention first. He sheds so much hair on our carpets that it clogs the vacuum cleaner. He leaves his daily marks in the yard and eats the cat food every chance he gets. I really wish Eddy didn't do these things, but we recognize that it's all just part of being a dog. We can't condemn or resent Eddy, because we love him and we appreciate the magnificent service he provides to our blind daughter: companionship, protection, and mobility.

A long time ago we had another *very* natural dog named Kippy. Our children loved Kippy, but I must confess that my feelings bordered on hatred. You see, Kippy resented backyard captivity, and he dug more holes under the fence than you could count. I really resented that dog for all the time he required of me trying to plug the holes, find him, and bring him back—and, I suppose, because he always outwitted me. Before the contest was over, the ground along our entire 200 feet of fence was lined with 2 x 8 planks and cinder blocks to cover his escape holes.

I now understand why I resented Kippy so much. It's because, at that time in my life, I didn't understand that he was just being "a natural dog." And so I thought God must feel the same way about *me* and all the things in *my* life that he didn't like or approve of. But now I know better. Now I know that God doesn't judge or resent any of his children for being who we are and what we are. Yes, it's true that "the natural man is an enemy to God" (See Mosiah 3:19) and that he "cannot look upon sin with the least degree of allowance" (D&C 1:31), but not because he resents the way we are. It's because if we *stay* that way, we'll never get to be with him again. He therefore spends all his time lovingly trying to help us escape being a "natural man" kind of person so that we may become a Christlike person and enjoy his divine fellowship. (We'll discuss this problem and opportunity in chapters 8-11.)

As far as I know, Eddy and all the rest of the natural dogs are going to be that way forever. And that's okay. I guess that's the way they were created—to be dogs. They do what dogs do. But people are different. We are God's children. We're meant to be better than we are in this fallen, natural state.

Thankfully, I've learned that it's stupid to resent dogs for being themselves. And I've also learned that it isn't wise to condemn *ourselves* (or others) for being a "natural man." It isn't our fault. We inherited the natural-man condition just by being born into this fallen world. To grow spiritually and overcome this lesser state, we must accept the fact that temporarily, for this life, our physical bodies, the houses in which our spirits live, are like Eddy, the natural dog. They have desires and traits we don't like and wish we didn't have. But we can learn to overcome these imperfections and rise above them. That's why the Lord has revealed tens of thousands of verses about the various parts of our bodies, teaching us the things he wants us *to do* and *not* to do with our body members so that we can achieve victory over the flesh and not become slaves to it.

We fought a war for this privilege. Let's learn how to make the most of it.

WHY WE SHOULD FOLLOW INSTRUCTIONS

"Your sins have withholden good things from you" (Jer. 5:25).

Driving along the freeway one day, I passed an interesting truck. It was constructed with several tapered bins so that whatever it was carrying could be drained out the spouts at the bottom. Apparently the bins needed a little vibration to help drain the product, because on the side of each bin was a large warning sign: "Do not beat on this bin with anything but a rubber hammer."

That warning surprised me. Who would be stupid enough to beat on a metal bin with a sledgehammer and not expect to leave dents? It seems obvious, doesn't it? And yet how many of us mistreat our bodies, leaving spiritual dents because we don't heed the warning signs our Heavenly Father has placed in the scriptures? Heavenly Father wants us to come out of this life victorious over the carnal desires of our flesh, not deformed with spiritual dents and passions that control our lives and keep his Spirit from us.

I once purchased a pint of milk and tried unsuccessfully to open the wrong side of the carton. Frustrated, I looked at the carton to see what was wrong and saw those familiar instructions: "Open other side." I decided to repent of my error. Turning the carton, I found the guiding words: "Open here." I did so and quickly obtained access to the milk.

Unfortunately, many people struggle their way through the issues of life refusing to heed God's instructions, shouting, "No! Don't tell me what to do. It's *my* body and I'll do what I want with it." There are many in the world who want no part of God or his rules of happiness. "Behold, they do not desire that the Lord their God who hath created them should rule and reign over them" (Hel. 12:6). "Therefore they say unto God, Depart from us; for we desire not the knowledge of thy ways" (Job 21:14).

Of course, we have the agency to abuse our bodies and do things the hard way if we insist. But how much wiser it is to follow the loving counsel of our Heavenly Father, whose every word is calculated to help us get the most joy and growth out of mortality. The scriptures are filled with divine instructions for the use of our bodies. We'll have greater success and joy in our lives when we do our best to follow instructions and allow ourselves time to grow as we work through the inevitable mistakes that are part of the tutoring we came here to experience.

Everything we do with (or to) our physical bodies is crucially important, because each act and choice contributes to thought and habit patterns that can last into eternity. "For that same spirit which doth possess your bodies at the time that ye go out of this life, that same spirit will have power to possess your body in that eternal world" (Alma 34:34). Every choice and every act adds strength or weakness to our characters. *Where* we'll live in the next life, *how* we'll live, and *who* we're allowed to spend eternity with depends, to a very large measure, on how we deal with our desires and passions in *this life*, for "in the last day it shall be restored unto him according to his deeds" (Alma 42:27). "Therefore, remember, O man, for *all thy doings* thou shalt be brought into judgment" (1 Ne. 10:20).

The Lord has warned that the day of judgment will include every secret desire and act we have done with our bodies, for "there is nothing secret save it shall be revealed; there is no work of darkness save it shall be made manifest in the light" (2 Ne. 30:17). Bruce R. McConkie explained one way our thought patterns and values are recorded and could be revealed in the judgment:

> In a real though figurative sense, the book of life is the record of the acts of men as such record is written in their own bodies. It is the record engraven on the very bones, sinews, and flesh of the mortal body. That is, every thought, word, and deed has an affect on the human body; all these leave their marks, marks which can be read by Him who is Eternal as easily as the words in a book can be read.
>
> When the book of life is opened in the day of judgment (Rev. 20:12-15), men's bodies will show what law they have lived. The Great Judge will then read the record of the book of their lives; the account of their obedience or disobedience will be written in their bodies *(Mormon Doctrine* [Salt Lake City: Bookcraft, Inc., 1966], p. 97).

"FEARFULLY AND WONDERFULLY MADE"

One reason I wrote this book is to help us live more spiritual, victorious lives. One way to do that is to gain a greater respect for the majesty of our wondrous tabernacle. Let's see if we can gain a greater reverence and appreciation for our physical bodies by considering some of the things about our physical houses we sometimes take for granted.

For example, only 40% of the body is made up of solid materials; the rest is water! In an average-sized adult man, that 60% of water makes up about 84 pints, which is more than ten gallons! In this day of microchips and electronic circuits, one wonders how ten gallons of water could possibly operate as a human body. Of course, the body is much more than water. The human body contains hundreds of different chemicals. In fact, the body could be described as a chemical machine, because there are billions of fragile chemical reactions which take place every second to sustain our life and give us energy.

Our bodies contain about 75 trillion cells—the building blocks of life. These tiny cells differ vastly in their shape, size, and function. About 30 trillion of these are red blood cells, which circulate throughout the body twice a minute through a network of 6,000 miles of blood vessels, delivering oxygen and nutrients and carrying away waste materials. If you could lay all your blood vessels end-to-end, they'd stretch across the United States from coast to coast and back again. It's been estimated that almost 200 billion of these cells become ineffectual and die every hour. But in a healthy body, these dying, worn-out cells are replaced by new ones without our even being aware of the process.

The human body also contains more than 200 bones, which are linked at more than 100 movable joints. (Imagine a machine with that many flexible points!) Our skeletal structure is stronger, pound for pound, than steel. In fact, if the skeleton were built of steel to an equivalent strength, it would have to weigh about five times as much as it does. And bones, like the rest of the body, are self-repairing.

Whether you're running, blinking, or just standing still, you're using muscles. Thirty to fifty percent of our body weight is muscle. Many of our muscles are classed as "involuntary" because they control things like the heartbeat, breathing, or the movement of food and fluids through the body. We have about 650 "voluntary" muscles, which are under our conscious control. Once we master a motor skill like walking, running, dancing or skiing, the brain automatically controls our muscles. We usually take this for granted until we try to learn a new skill like typing or playing an instrument. Then the task of coordinating more than six hundred pulling devices simultaneously helps us to appreciate the complexity of this human machine.

Our bodies are covered with about twenty square feet of skin, which not only holds our "insides" in and gives shape to our body but also keeps germs and the rays of the sun from damaging the interior of our bodies. The skin is tough but delicate; it looks simple but is very complex. Within a patch of skin the size of a pea are about fifteen oil glands, which keep it soft and flexible, and about 100 sweat glands.

The skin helps control our body temperature by shunting the warm blood into deeper tissues when we're cold and sweating when we're too warm. This complex and automatic temperature control is accomplished by some two-and-a-half million sweat glands. If you could lay them all end-to-end, they would stretch about thirty miles! An area the size of a pea contains about fifteen hairs and 230 sensor nerve endings, which detect touch or pressure and inform us about the environment's effects on our bodies. It also contains half-a-million dying cells. Every day we rub off millions of dead skin cells. (In fact, a large amount of the dust you find in your home is actually flakes of dead skin.)

Our lungs, which make it possible to provide oxygen to the blood and remove the carbon dioxide, contain about 700 million alveoli (tiny air sacs) where these gases are exchanged between the lungs and the blood. The combined surface area of these alveoli is some 84 square yards—almost the size of a tennis court. The average number of breaths per minute is sixteen. This equals about seventy-five gallons of air exchanged every minute, automatically, without our conscious control or concern. As King Benjamin taught, God not only created our bodies and gave us life, but "is preserving [us] from day to day, by lending [us] breath, that [we] may live and move and do according to [our] own will, and even supporting [us] from one moment to another" (Mosiah 2:21).

As this air is drawn in through the nostrils, it swirls about the nasal cavity before passing down the windpipe to the lungs. High up on each side of the nasal cavity is an area about one inch square where smells are sensed. These two sensory sites have millions of nerve endings, shaped like tiny bowling pins, which react to particles in the air that carry odor-emitting substances. Humans have about 16 million olfactory cells in the nose. (Rabbits have 100 million! Perhaps that's why their noses twitch all the time?) Most of us can detect about

4,000 different odors, but an expert's nose can be trained to detect as many as 10,000 different smells. The nose's self-defense mechanism against inappropriate intruders can expel air during a sneeze at about 100 miles per hour!

The marvel is that all this happens *automatically*, regulated by the intelligence God has placed within the brain, glands, and organs, so that *they* run the system for *us*—the spirit entities within—so that we can occupy our conscious thoughts and efforts with our journey of eternal progression.

Your eyelids blink automatically every two to ten seconds. That's between 5,000 and 28,000 blinks a day to keep the exposed surface of the eyeball moist and protected from bacteria. Aren't you glad you don't have to remember to do that consciously? Cut your finger and it heals itself. Provide this incredible masterpiece with proper nourishment and rest, and it's largely self-sustaining. Put food in your stomach, and your body knows how to digest it and convert it to energy. Decide, with your mind, where you want to go, and your body will take you there with almost no conscious thought on your part.

This body that we wear is a marvel of engineering unequaled by any other creation we know of. Truly, "I am fearfully and wonderfully made" (Ps. 139:14). The vast complexity of intricate and interwoven details of the design and function of our physical bodies were all conceived and planned by God before we were created. "Thine eyes did see my substance, yet being unperfect; and in thy book all my members were written, which in continuance were fashioned, when as yet there was none of them" (Ps. 139:16). "But now hath God set the members every one of them in the body, as it hath pleased him" (1 Cor. 12:18).

OUR BODIES AS TEMPLES

The Lord has pleaded with us to have reverence, respect, and appreciation for our mortal bodies and to treat them like temples. "Man is the tabernacle of God, even temples" (D&C 93:35). This means that the physical body is not given to *us* exclusively as a covering for *our* spirits. Our stewardship is to keep it clean, pure, and holy like a temple so that it will also be a fit habitation for the visitation of the Holy Ghost and the Spirit of the Lord as well. "The temple of God is holy, *which temple ye are*" (1 Cor. 3:17). "Know ye not that ye are the temple of God," asked Paul, "and that the Spirit of God dwelleth in you?" (1 Cor. 3:16). "What? know ye not that your body is the temple of the Holy Ghost which is [supposed to be] in you, which ye have of God, and ye are not your own?" (1 Cor. 6:19).

Jesus made his first visit to the temple in Jerusalem when he was only twelve years old. From that time forward, he loved to visit his Father's house. But when he found it defiled with commercial ventures that polluted the sacredness of that place of worship, he was greatly displeased and drove the offenders from the temple. (See Matt. 21:12-13.) Jesus cleansed the temple in Jerusalem so that he could feel comfortable there, and so that the people who came to worship could feel his Heavenly Father's Spirit. How eagerly we should strive to keep our temple-body clean and pure so that he feels comfortable to visit us! "The Lord hath said *he dwelleth not in unholy temples*, but in the hearts of righteous doth he dwell," but "*if it be defiled I will not come into it*, and my glory shall not be there; for I will not come into unholy temples" (Alma 34:36; D&C 97:17).

Our duty is to keep our bodies pure and clean and obedient, no matter how much we must sacrifice to do so. The principle of keeping our homes, our bodies, and our personal lives clean and holy in preparation for the Lord's visit goes clear back to the Old Testament. "For the Lord thy God walketh in the midst of thy camp . . . therefore shall thy camp be holy: *that he see no unclean thing in thee, and turn away from thee*" (Deut. 23:14). "And he doth not dwell in unholy temples; neither can filthiness or anything which is unclean be received into the kingdom of God" (Alma 7:21). "I beseech you therefore, brethren, by the mercies of God, that ye present your bodies a living sacrifice, holy, acceptable unto God, *which is your reasonable service*" (Rom. 12:1).

I suppose dogs are happy being dogs. But people will never be truly happy until they rise above the carnal, natural man to enjoy the spiritual life God intended for us. And that's what this book is all about. If your car develops a mechanical defect, you don't condemn *yourself*. You take it to a repair specialist and then get on with your life. To get ourselves repaired, we *must* come to Christ, for you and I cannot change our human nature or our natural-man traits by ourselves any more than Eddy or Kippy can stop being a natural dog.

Someday—if we live for it by doing the best we can during this life to conquer, or at least control, the unworthy desires of our fallen flesh—if we allow Christ to change us and make us new—we'll be given perfected bodies that won't be at war with the desires of our spirits. We'll have peace and joy for eternity. But until then, let's not waste our lives judging and resenting dogs for being dogs, or ourselves for temporarily living in fallen bodies.

Chapter Two

THE EYES

"The way is prepared, and if we will look we may live forever" (Alma 37:46).

Connecting our brains to the outside world, our eyes are among the most important and sacred parts of our bodies. The greatest flow of information to our brains comes through our eyes, which receive about 75% of all perceptions of our environment. The eyes have been compared to biological movie cameras, but eyes are far more than mere optical receivers. Just as we use a keyboard to input data into a computer, we use our eyes to feed data into our computer-brains. When you add all the verses about eyes, looking, seeing, and blindness, the scriptures contain almost 3,000 revelations on the things God has asked us *to do* with our eyes and the things he has asked us *not to do.*

Our incredible eyes have been given such marvelous adjustment capabilities that you can gaze in wonder at the beauty of the moon a quarter of a million miles away, look down at your hand, and instantly refocus on the moon again. At the back of the eye is the retina, not much larger than a postage stamp and about the same thickness. This is a light-sensitive layer of 130 million cells that collect light and images. Inside each of these light-sensitive cells are up to 100 million molecules of photo pigment. The human eye is so sensitive that on a clear night, it can actually detect a match being lit fifty miles away!

The question for our spiritual anatomy is: Where do we focus our eyes and attention each day? How sensitive are our eyes in detecting the spiritual needs of those around us who need more of the light of Christ?

As our eyes collect information about the world around us, these delicate and sensitive devices convert light energy into complex patterns of nerve signals. Every second, the eyes transmit millions of these nerve signals through the optic nerve fibers to the visual cortex at the back of the brain, where visual images are recreated and evaluated for relevance and processed for appropriate responses. As the primary source of data to the brain, our eyes have enormous power to influence our thoughts, attitudes, and choices.

HOW GOD USES HIS EYES

One of the most incredible concepts revealed in scripture is what God does with his eyes. With so many galaxies and worlds to monitor and care for that they are described as "without number," he still watches and ponders the lives of each of his children. "For his eyes are upon the ways of man, and he seeth all his goings" (Job 34:21) and "from the place of his habitation he looketh upon *all* the inhabitants of the earth" (Ps. 33:14).

This earth upon which we live, this *school* where you and I are determining our eternal destinies, is "a land which the Lord thy God careth for: the eyes of the Lord thy God are always upon it, from the beginning of the year even unto the end of the year" (Deut. 11:12).

It's important to realize that God's viewing of Earth is not of the planet or mankind in mass, but of *each* person, whom he loves *individually*. Not only does he observe our daily actions and choices, but he also ponders the best circumstances to bring into our lives so that we may learn and grow. "For the ways of [each] man are before the eyes of the Lord, and he pondereth all his goings" (Prov. 5:21).

> And Ammon said unto him: The heavens is a place where God dwells and all his holy angels.
> And King Lamoni said: Is it above the earth?
> And Ammon said: Yea, and *he looketh down upon all the children of men*; and he knows all the thoughts and intents of the heart (Alma 18:30-32).

I'm fond of two Old Testament scriptures (even though they have some obvious translation problems) as they describe "the eyes of the Lord, which run to and fro through the whole earth" (Zech. 4:10). We understand, of course, that his eyes are not literally running, but rather are *scanning*, searching, and observing. The second scripture tells us that one reason God observes our lives so carefully is to find opportunities to reward our faithfulness. "For the eyes of the Lord run to and fro throughout the whole earth, to shew himself strong in the behalf of them whose heart is perfect toward him" (2 Chr. 16:9).

"God looked down from heaven upon the children of men, to see if there were any that did understand, that did seek God" (Ps. 53:2). Several times in modern scripture, Jesus Christ demonstrated the truth of these verses by singling out specific individuals and stating to them, through the prophet Joseph, that he had known them personally and watched and appreciated their service in building the kingdom. For example, to James Covill, he said: "And now, behold, I say unto you, my servant James, I have looked upon thy works and I know thee" (D&C 39:7). Likewise, to Sidney Rigdon, he said: "Behold, verily, verily, I say unto my servant Sidney, I have looked upon thee and thy works" (D&C 35:3).

When he looks upon your works, and mine, is he pleased or concerned? After observing Isaac Galland's faithful service in the Church, the Savior said, "I, the Lord, love him for the work he hath done, and will forgive all his sins" (D&C 124:78). Not only are his eyes "upon the ways of man, and he seeth all his goings" (Job 34:21), but "he withdraweth not his eyes from the righteous" (Job 36:7). "For thine eyes are *open upon all the ways of the sons of men* to give every one according to his ways, and according to the fruit of his doings" (Jer. 32:19).

Elder Paul H. Dunn has given testimony that "God the Father has not forgotten us here in mortality. He has not removed himself to a far corner of the universe to watch our antics with indifference." He said, "Many people believe that he's done just that. They can't believe that he could create a universe, people a world with billions of souls, and still care a whit what happens to a single individual with his small concerns. They can't believe that they're that important to anyone, let alone to the Creator of it all. May I tell you that I know that God lives, that he cares, and that he knows each one of us individually by name" (*Ensign*, May

1979, p.8). "Now my brethren, we see that God is mindful of every people, whatsoever land they may be in; yea, he numbereth his people, and his bowels of mercy are over all the earth" (Alma 26:37).

Just a final note about God's eyes. You may be interested to know that the expression "the apple of his eye" is Biblical. It's found in Deut. 32:10; Ps. 17:8; and Zech. 2:8.

WHAT GOD WANTS US *TO DO* WITH OUR EYES

With all the things we have to look at in this world, the four specific places and priorities Christ has established for us to focus our vision upon are:

1. Observing nature to see the Creator.
2. Visualizing the atonement to see Christ as our Savior.
3. Looking to Christ for the help we need.
4. Having eyes that are single and focused.

After we evaluate these four concepts, we'll consider some of the things the Lord has asked us *not* to do with our eyes.

OBSERVING NATURE

One of the simplest and most enjoyable things God has asked us to notice and appreciate is the beauty of his creation. At each stage of creation of this physical earth, God pronounced it good. He has told us that "all things which come of the earth, in the season thereof, are made for the benefit and the use of man, both to *please the eye and to gladden the heart*" (D&C 59:18).

Jesus said that "it pleaseth God that he hath given all these things unto man" (D&C 59:20), but warned that as merciful and patient as God is, it's possible to offend him when we "confess not his hand in all things" (D&C 59:21). I wonder if that principle might extend to failing to *notice* his hand in all things? We risk offending our creator when we rush through life so preoccupied that we fail to notice the beauties of his handiwork, which he manifests not only to "*please the eye and to gladden the heart*," but to testify of his love and glory. "The heavens declare the glory of God; and the firmament sheweth his handiwork" (Ps. 19:1).

Isaiah counseled that we should train ourselves to appreciate the beauties of nature. "Lift up your eyes on high, and behold who hath created these things" (Isa. 40:26). James E. Faust gave the same advice. "Develop an appreciation for the great gifts of God as found in nature," he said, "the beauty of the seasons, the leaves, the flowers, the birds, the animals" (*Ensign*, 1998, p.4).

One of our spiritual challenges in mortality is to remember in our daily lives *who* we are and *why* we're here. Every Sunday we devote special time during the sacrament service to remembering and renewing our covenants with Jesus Christ—and *promising to remember him* throughout our daily activities. We can carry this important principle of remembering him into our daily lives by noticing and appreciating the active hand of God in the ever-changing beauties of nature.

I can testify that when I take the time to see God's hand in a beautiful sunrise or sunset, in the sweet song of a bird, or the color and fragrance of a flower, I find myself caring less about the troubles of the world and more about worshiping and honoring my Creator.

VISUALIZING THE ATONEMENT

Nothing in the gospel eclipses the importance of the sacrifice and atonement of Jesus Christ. And yet how easy it is to take it for granted, to ignore it, or fail to appreciate its magnitude.

Our brains make it possible for us to *visualize*, or *see*, mental images of things not in our physical field of view. In this manner we can *see* things both past and present. The scriptures counsel us to use this ability to visualize the details of the atonement so that, by becoming real and vivid in our minds, these events gain the power to affect our feelings and commitment. "Wherefore," said Jacob, "we would to God that . . . all men would believe in Christ, and *view his death*, and suffer his cross and bear the shame of the world" (Jacob 1:8).

What did Christ mean when he commanded us to "*look* unto me in every thought . . . [and to] *behold* the wounds which pierced my side, and also the prints of the nails in my hands and feet. . . ."? (D&C 6:36-37). How can one "view" and "behold" these things which took place so long ago except by devoting time to visualization? The dictionary defines the verb "visualize" as the act of making something visible by forming a mental image.

When the scriptures command us to view these things, it means that we're to study the accounts and to mentally and emotionally place ourselves in Gethsemane and at the cross so that we can visualize his agonizing ordeal. His suffering needs to become so real to us that we can actually see the blood oozing from every pore of his tortured body and know that it was done because of *our* sins and his love for *us*. It means that we should try to hear his groans of agony as he took upon himself the effects of *our* own sins—*our* pain, *our* guilt and shame, *our* punishments—the same as if he had committed the sins himself. (See Isa. 53:4–8.)

We should also be able to view his sufferings on the cross with such reality that we can imagine the splinters of rough wood piercing his quivering flesh as he is cruelly thrust down upon the cross, his back still raw from the Roman scourging. We should be able to hear the sound of the mallet as it drives rough, rusty spikes though the flesh of his hands and feet. The Savior's sacrifice on the cross should become so real and vivid to us that we can see and hear the agonizing tearing of his flesh as the heavy cross is lifted, then cruelly dropped into the hole with a heartrending thud. To obtain full benefit from the atonement, we need to literally and vividly view his death and behold the wounds as he hangs there, exhausted, misunderstood, hated, and rejected of men because he loved us. "Is it nothing to you, all ye that pass by? *behold, and see* if there be any sorrow like unto my sorrow, which is done unto me, wherewith the Lord hath afflicted me in the day of his fierce anger" (Lam. 1:12).

As Isaiah described the atonement and how Christ would suffer the consequences of our sins and other burdens so that we wouldn't have to, he said, "and with his stripes we are healed" (Isa. 53:5). The Savior's "stripes" not only refer to the lashes of the Roman whip upon his bare back, but also symbolize all that he suffered and bore on our behalf. When we train ourselves to see this infinite sacrifice as it really happened, both *for* us and *because* of us, our hearts are changed, our behavior is changed and elevated, and we're more cautious about the things we allow our eyes to behold. We receive the Savior's peace as we are truly healed of our spiritual burdens.

LOOKING TO CHRIST FOR SPIRITUAL RESOURCES

An interesting story in the New Testament concerns Zacchaeus, a short, rich man in Jericho. As the Savior passed through the town he was, as always, followed by a crowd of people pressing close to hear his words. Zacchaeus also "sought to see Jesus who he was; and could not [see] for the press, because he was little of stature" (Luke 19:3). But Zacchaeus was a man with determination. Using his feet in a wise manner, "he ran before, *and climbed up into a sycamore tree to see him*: for he was to pass that way. And when Jesus came to the place, he looked up, and saw him, and said unto him, Zacchaeus, make haste, and come down; for to day I must abide at thy house" (Luke 19:4-5).

We're all "short" on the spiritual qualities it takes to see and to fellowship with our Savior. The question is, what haste and effort do we make to be close to him and to see his influence in our lives? Since he is revealed in the scriptures, it isn't necessary for us to climb trees. We need only open the pages and ponder the revelations. And when we do, we may be sure that he'll be just as aware and rewarding of our efforts as he was of Zacchaeus'.

After a lifetime of growing up and waiting for the time that we can do it all by ourselves, it's natural to look to ourselves and our own resources to solve our problems. But now that we're grown and facing temptations, obstacles, and trying circumstances, Christ is not only asking us to *look* unto him "in every thought" (D&C 6:36), but also to learn to *relate* to him as a little child who knows he is dependent on someone else. "Except ye be converted, and become as little children, ye shall not enter into the kingdom of heaven" (Matt. 18:3). "Therefore, whoso repenteth and cometh unto me as a little child, him will I receive, for of such is the kingdom of God" (3 Ne. 9:22).

It's true that we have many resources in times of trouble. The world is filled with books, self-improvement and behavioral modification programs, and counselors to whom we can look for help in our needs. But unless we live close to the Savior, looking to him for guidance, strength, and help, our problems will continue to be difficult. While those sources are certainly beneficial, only when we look to Jesus Christ for the true light and resources we need will we achieve the peace, success, and abundance in life that he meant us to have. Isaiah warned that sorrow would befall those who look elsewhere, trusting in the more obvious arm of flesh, instead of relying on Jesus Christ:

> Wo to them that go down to Egypt for help; and stay on horses, and trust in chariots, because they are many; and in horsemen, because they are very strong; *but they look not unto the Holy One of Israel,* neither seek the Lord! (Isa. 31:1; 45:21-22).

> And we talk of Christ, we rejoice in Christ, we preach of Christ, we prophesy of Christ, and we write according to our prophecies, *that our children may know to what source they may look* for a remission of their sins (2 Ne. 5:26).

Do not misunderstand. While it's both desirable and necessary to look to our own strengths and abilities and to do as much as we can, it's also imperative to recognize that no one, by his own efforts alone, can ever do enough to live the obedient life that will result in exaltation. It seems noble and valiant, but it's Satan's lie that we must do it all on our own, that we mustn't look to God for help to obey the very commandments which he gave us.

> We must do all that we can do. We must extend ourselves to the limit; we must stretch and bend the soul to its extremities. In the final analysis, however—at least when dealing with matters pertaining to spiritual growth and progression—it is not possible to "pull ourselves up by our own bootstraps," nor is it healthy to presume we can (Robert L. Millet, *By Grace Are We Saved* [Salt Lake City: Bookcraft, 1989], p.100).

As we learn to look more to the Lord and his divine resources than to ourselves, one of the things we symbolically do with our eyes is to view ourselves: not as if looking in a mirror, but at the image of our inner, real selves. The Lord is there to help us see what we need to see in order to grow closer to him, for he has promised divine help in this self-analysis. "And if men come unto me I will show unto them their weakness" (Ether 12:27). He does this to kindly show us our own defects so that we can come to him more fully. There is grave danger in trusting the way we see ourselves instead of the way the Lord sees us, for our opinions are seldom objective. Nor is it likely that we can see ourselves as clearly as God does. "Every way of a man is right in his own eyes: but the Lord pondereth the hearts" (Prov. 21:2). "All the ways of a man are clean in his own eyes: but the Lord weigheth the spirits" (Prov. 16:2).

While "the Lord cannot look upon sin with the least degree of allowance" (D&C 1:31; Alma 45:16), the normal tendency of the natural man is to rationalize his sins instead of making himself repent and grow. We're masters at justification. Just as Lot's wife justified looking back (and suffered the terrible consequences), we often justify the things we do because "everybody does it" or for some other nonsensical reason. The type of entertainment we view or participate in is a good example. It's easy to deceive ourselves. "There is a generation that are pure in their own eyes, and yet is not washed from their filthiness" (Prov. 30:12).

We may not think we're proud until we face the reality that we may not be as wise or as wonderful as we appear to our own view. Let's remember the counsel of the prophets: "The way of a fool is right in his own eyes" (Prov. 12:15), but "woe unto them that are wise in their own eyes, and prudent in their own sight!" (Isa. 5:21).

As we ponder the atonement and our relationship with Christ during the sacrament each week, we must each consider our dependence on him for the ability to keep our covenants and to use our faculties in ways that will honor him and diminish our natural-man tendencies. We must ponder intently to see ourselves through *his* eyes instead of our own.

HAVING SINGLE EYES

Two concepts are taught in scripture in regard to single eyes: the first has to do with our commitment and dedication to the Lord's work, and the second has to do with whether our lives are filled with the Light of Christ or the darkness of Satan.

Dictionaries define the word "single" as being "wholly attentive, not divided and having the same or consistent application for all." Therefore, if we want to qualify ourselves for effective service in our callings, for example, we must not only develop faith, hope, charity, and love as attributes, but also commit ourselves "with an eye single to the glory of God" (D&C 4:5). That is to say, we must be focused on eternal perspectives; we must care more about the Lord's affairs than we do about our own; and we should be willing to sacrifice personal time and effort to further his purposes. It's called seeking first the kingdom of God.

Great rewards come to those who commit themselves to the Lord's work with "an eye single to his glory"; great perils come to those who will not. "The light of the body is in the eye," said Christ. "If therefore thine eye be single, thy whole body shall be full of light. But if thine eye be evil, thy whole body shall be full of darkness. If therefore the light that is in thee be darkness, how great is that darkness!" (Matt. 6:22-23; also 3 Ne. 13:22-23). "And if your eye be single to my glory, your whole bodies shall be filled with light, and there shall be no darkness in you; and that body which is filled with light comprehendeth all things" (D&C 88:67).

THINGS THE LORD HAS ASKED US *NOT TO DO* WITH OUR EYES

The two major things the Lord has asked us *not* to do with our eyes are to use them to judge others or to lust after their bodies.

JUDGING OTHERS

One trait of the natural man is his eagerness to judge how others are living *their* lives so he can make a more favorable judgment of his own life. We're far more eager to observe *their* follies than we are to see our own. The scriptures refer to this part of human nature as hypocrisy and advise us to tend to our own business before we so eagerly seek to tend to that of others.

> Either how canst thou say to thy brother, Brother, let me pull out the mote that is in thine eye, *when thou thyself beholdest not the beam that is in thine own eye?* Thou hypocrite, cast out first the beam out of thine own eye, and then shalt thou see clearly to pull out the mote that is in thy brother's eye (Luke 6:42).

> And why beholdest thou the mote that is in thy brother's eye, but considerest not the beam that is in thine own eye? (Matt. 7:3).

Paul said, "Thou art inexcusable, O man, whosoever thou art that judgest: *for wherein thou judgest another, thou condemnest thyself;* for thou that judgest doest the same things" (Rom. 2:1). If that condemnation seems harsh, be aware that Jesus has repeated it in our own day, saying, "Those who cry transgressions do it because they are the servants of sin, and are the children of disobedience themselves" (D&C 121:17). Since "God sent not his Son into the world to condemn the world" (John 3:17), it's obvious that we should be careful about using our eyes in judgment or criticism of others.

EYES OF LUST

We know that there are many natural, but unworthy and inappropriate, desires that people experience, not because they are wicked, but simply because their spirits live in mortal bodies of flesh. Mortal, fallen flesh has its own desires entirely independent of our personal will or choice. (See 2 Ne. 10:24; Rom. 6:12.) Righteous people can spend their entire lives learning to conquer these desires, but in doing so it's important to realize that as fallen beings, we struggle against *two* distinct lusts: the lust of the *flesh* and the lust of the *eyes*. "For all that is in the world, the lust of the flesh, *and* the lust of the eyes, and the pride of life, is not of the Father, but is of the world" (1 John 2:16).

Failing to conquer this deadly lust of the eyes can affect everything in our lives. One man, for example, can look at a beautiful woman and see her as a lovely daughter of God with extraordinary features. Another man, looking at the same woman with eyes of lust, sees not the beauty of her creation or her person, but merely an object of lust, a stimulation for fantasy and imagined pleasures. "Having eyes full of adultery [they] cannot cease from sin" (2 Peter 2:14).

The scriptures are replete with warnings about the dangers of using our eyes as keyboards to input improper data and images into our minds, especially pornographic images. One of the challenges a natural man faces in learning to control eyes of lust is his preoccupation with the nude form. There seems to be a natural fascination for it, but the provocative nudity portrayed in pornography defiles the private, intimate, and sacred nature of sexuality by putting it on public display. Prophets who counsel us not to use our eyes to view nude bodies are not implying that there is something evil about nakedness. Nudity is not evil of *itself*, but is sacred and reserved for the private relationship between husbands and wives.

The scriptures teach that no one has the right to view another person's nudity unless invited to do so within the marital relationship.

When a man allows his eyes to roam over the body of a woman who is not his wife, his mental lusting is a violation of the loyalty he owes his wife. The scriptures advise men to reserve their sexual desires exclusively for their wives, to "be thou ravished always with *her* love," and to "let *her* breasts satisfy thee at all times" (Prov. 5:18-19). Here is the blunt statement of principle about viewing nakedness outside of marriage, as declared by the Lord himself:

> And if a man shall take his sister, his father's daughter, or his mother's daughter, and see her nakedness, and she see his nakedness; *it is a wicked thing;* and they shall be cut off in the sight of their people: he hath uncovered his sister's nakedness; *he shall bear his iniquity* (Lev. 20:17).

Does that warning apply today? Every daughter of Adam and Eve is a sister. Our bodies are created in the image of God. He created them to be attractive to each other. "It's perfectly natural to be attracted to a good-looking young woman. It's the way God planned that it should be. Satan knows this, however, and will ever so quietly sneak bad thoughts into your mind" (Editorial, *The New Era*, May 1989, p. 17).

There's a fine line, a very dangerous line, between the natural feelings of appreciation for the beauty of the bodies God created and the danger of allowing those righteous attractions to progress into lustful desires. "Lust not after her beauty in thine heart" (Prov. 6:25). As one mission president counseled his elders, "There's nothing wrong with thinking, 'She's sure pretty.' But let the thought stop there"(*The New Era*, May 1989, p. 18).

King David failed to heed the warning against lusting in his heart. "And it came to pass in an eveningtide, that David arose from off his bed, and walked upon the roof of the king's house: and from the roof he saw a woman washing herself; and the woman was very beautiful to look upon" (2 Sam. 11:2).

Nothing wicked had occurred at this point. The woman wasn't being provocative. She thought she had privacy in the dark of night. David wasn't prowling the neighborhood peeking in windows or seeking thrills. He was simply having trouble sleeping and was relaxing in his private domain. Accidentally he discovered a woman of beauty within his line of vision. It wasn't wicked to *see* her, but it led to wickedness to *keep* seeing her. It stimulated desires for another man's wife. It fanned the normal, mortal desires that are natural to fallen man.

If David had turned his head and disconnected the input, he could have turned his thoughts and controlled his actions. But he did not turn. He stared and drank it in until his mind was aflame with desire and possibilities. Then he used his kingly powers to bring about the fulfillment of those unrighteous desires and has, perhaps, paid for the momentary lustful pleasure with his eternal salvation.

The dilemma we all face as mortal, fallen people is that *"the eyes of man are never satisfied"* (Prov. 27:20). Unlike eating or drinking, which temporarily *satisfies* the craving, viewing pornography *does not* satisfy one's curiosity or lustful desire, but feeds it and creates an *ever increasing hunger* for more and more explicit images.

David committed adultery with this woman right there on his roof before they ever came together physically, for "whosoever looketh on a woman to lust after her hath committed adultery with her already in his heart" (Matt. 5:28; 3 Ne. 12:28). We came to earth to learn to build relationships with *people*, not to waste our lives in preoccupation with their bodies. When we substitute fascination with the beauty of the *outward body* for respect for the person inside, we become vulnerable to great danger.

There are at least two major reasons why mental adultery is so perilous. First, mental adultery plants the seeds we eventually harvest in future acts; and second, mental adultery causes loss of the Spirit and therefore, loss of our spiritual battles.

MENTAL ADULTERY PLANTS THE SEEDS OF FUTURE ACTS

The first danger of pornography and mental adultery is the obvious truth that *physical* adultery is always preceded by *mental* adultery. The visual inputs we choose (or allow) give birth to thoughts and desires and then those thoughts give birth to our actions.

One of Satan's widely believed lies is that it doesn't hurt to *look* at nudity as long as we don't get involved physically. But the nervous system can't tell the difference between a *real* experience and one that is vividly *imagined*. Anyone who has responded to an *imagined* fear knows that the body reacts just as it would had the emergency been genuine. Thus, the effect of pornographic memories, lust, and *mental* adultery are just as devastating to our progress and holiness as *physical* adultery. Perhaps that is why Jesus stressed that "thou shalt not . . . commit adultery . . . *nor do anything like unto it*" (D&C 59:6).

Satan uses pornography to attack virtue because he knows that if he can't tempt God's children to indulge in sexual sins *physically*, with pornography he can induce them to participate in those activities *mentally* and *emotionally*, thereby establishing mental habits and desires that plant the seeds that prepare the way for future acts of indulgence. Every chosen act is first preceded by thought, often by *patterns* and *habits* of thought. Each time we allow our keyboard eyes to feed pornographic material into our computer brains, we're planting indelible seeds of lust that can eventually grow into a harvest of compelling thoughts and immoral deeds as we eventually act out the things we have fantasized.

When the prophets counsel us to avoid R-rated movies and pornography, it's because it has been proven over and over that if we use our eyes to look at nudity long enough, we're going to end up acting out our lusts, because *the thoughts upon which we focus our attention are the blueprints of our future reality*. We always reap in our *physical lives* what we have been sowing in our *mental lives*—and when it comes to issues of morality and purity, those thoughts are largely controlled by visual input. We'll discuss the impact of our mental focus of attention on our outward behaviors in Chapter Nine, "The Mind."

The Savior said, "Whosoever *looketh* on a woman, to lust after her, *hath committed adultery with her already in his heart*" (Matt. 5:28). These familiar words show his intense concern about the seeds of desire and future actions planted by mental adultery. Can there be any doubt that looking longingly at pictures or videos of nude men or women is included in this warning?

A principle in Lot's escape from Sodom and Gomorrah relates to viewing pornography. The angels sent to rescue Lot's family gave specific instructions about their eyes: "Escape for thy life; *look not behind thee*, neither stay thou in all the plain; escape to the mountain, lest thou be consumed" (Gen.19:17). "Don't look."

You know what happened, though you may not understand it any better than I do. Lot's disobedient wife "*looked back* from behind him, *and she became a pillar of salt*" (Gen. 19:26). I can't tell you what was so sinful about Lot's wife simply looking back as the cities were destroyed. I don't know if her drastic punishment came because of some scientific principle involved in radiation from the fire and brimstone, or simply because of her disobedience in the face of a divine rescue. She lost her life in a way we may not understand fully. But the principle is that the representatives of God *gave her instructions* from her Heavenly Father about dangers that could come from specific use of her eyes, and her disobedience led to tragedy.

Today the prophets of God are giving *us* specific and detailed warnings about inputting pornographic images into our minds. We risk terrible consequences when we ignore them.

> We counsel you . . . not to pollute your minds with such degrading matter, *for the mind through which this filth passes is never the same afterwards.* Don't see R-rated movies or vulgar videos or participate in any entertainment that is immoral, suggestive, or pornographic (Ezra Taft Benson, Ensign, May 1986, p. 45).

> Recognize pornography for what it is—a vicious brew of slime and sleaze, *the partaking of which only leads to misery, degradation, and regret.* The Church expects you who have taken upon yourselves the name of the Lord Jesus Christ to walk in the sunlight of virtue and enjoy the strength, the freedom, the lift that come from so doing (Gordon B. Hinckley, *Ensign,* July 1997, p. 73).

How could we possibly consider this counsel about the use of our eyes to be any less serious than that which was given to Lot's wife? And because we ignore their counsel and disobey, we're losing our spiritual lives in epidemic numbers. We're not being turned into "pillars of salt," but we are being turned into slaves and sexual addicts, unable to control our lives or our thoughts. Families and lives are being destroyed and we're losing the Spirit of the Lord, all because we haven't learned to follow the counsel of our leaders. Jesus warned: "Remember Lot's wife" (Luke 17:32). Here's another warning from President Hinckley about the addictive dangers of pornography:

"Let virtue garnish thy thoughts unceasingly" (D&C 121:45). There is
so much of filth and lust and pornography in this world. We as Latter-day
Saints must rise above it and stand tall against it. *You can't afford to indulge
in it. You just cannot afford to indulge in it.* You have to keep it out of your
heart. *Like tobacco it's addictive, and it will destroy those who tamper with it.*
"Let virtue garnish thy thoughts unceasingly" *(Ensign,* Aug. 1997, pp. 6-7).

NOTE: For accounts of how two people recovered from addictions to
pornography, see the *Ensign,* September 1996, pp. 20-22, and Steven
A. Cramer's book, *The Worth of a Soul* [Springville, Utah: Cedar Fort
Inc., 1983].

MENTAL ADULTERY CAUSES LOSS OF THE SPIRIT

The second reason mental adultery is so perilous is that our computer brain can't run two programs at the same time. When we press
the enter key on the keyboard of our eyes, feeding in sexual, lustful
images, *we automatically press the delete key that drives the Holy Spirit
out.* Thus "he that looketh upon a woman to lust after her *shall* deny
the faith, and *shall not have the Spirit;* and if he repents not he *shall* be
cast out" (D&C 42:23). Does using the word "shall" three times in the
same sentence leave any room for doubt? And can there be any question that this warning applies equally to *pictures* of nudity? "And verily I say unto you, as I have said before, he that looketh on a woman
to lust after her, or if any shall commit adultery in their hearts, *they
shall not have the Spirit,* but shall deny the faith and shall fear" (D&C
63:16).

To win our battles with Satan, it's imperative that we have the
Lord's Spirit with us to expand and strengthen our efforts. *To lose the
Lord's Spirit is to lose our battles.* No matter how hard we try to rationalize our lustful eyes, the fact remains that holiness, virtue, and purity cannot coexist with deliberate filthiness and lust. Viewing pornography or indulging in sexual fantasies drives the Spirit from our minds
and hearts, *even if we do not act upon them,* for "the Spirit of the Lord
doth not dwell in unholy temples" (Hel. 4:24). Consider this sobering
warning by Elder Joseph B. Wirthlin: "Every *ounce* of pornography
and immoral entertainment will cause you to lose a *pound* of spirituality. And it will only take a few ounces of immorality to cause you to

lose all of your spiritual strength, for the Lord's spirit will not dwell in an unclean temple" (*The New Era*, May 1988, p.7).

When we lose the Spirit, everything that's important in our lives is diluted. Our family relationships are diminished; our enthusiasm for life and church service, our respect for sacred covenants, and our respect for consequences are all compromised.

The Lord warned that "if my people shall sow filthiness they shall reap the chaff thereof in the whirlwind; *and the effect thereof is poison*" (Mosiah 7:26). When we use our eyes to input pornographic nudity and acts of lust, we're literally poisoning our minds and spirits. As Elder Wirthlin also stated, "Pornography in all its forms . . . *constitutes a spiritual poison* that is addictive and destructive" (*The New Era*, May 1988, p. 7).

Note: For a cassette recording discussing the consequences of addiction to pornography and how to help someone overcome this problem, please refer to Steven Cramer's tape: *If Thine Eye Offend Thee* (American Fork, Utah: Covenant Communications, Inc., 1999).

THE URGENCY OF CONTROLLING OUR EYES

As we've seen, the relationship between our eyes and our minds is like that of a computer to a keyboard: Input = Output.

We all know how difficult it is to struggle with evil thoughts once they find their way into our minds. It's much simpler to disconnect the input than it is to expel the images or thoughts once they reach the contemplation stage inside the mind. If the act is to be prevented, the thought and contemplation must be stopped or its fulfillment becomes almost inevitable. Therefore, if we're struggling with immoral thoughts, one of the first steps to repentance is to disconnect the input. Then we have hope of mental and emotional cleansing, but not while we continue to feed the fires of passion.

Pornography could be likened to the atomic bomb of spiritual warfare. Just as fire and radiation from a nuclear explosion consumes the flesh of its victims, pornography consumes its victims with lust, blinding its users to eternal consequences. That's why controlling our eyes is not a principle with which we can indulge the slightest compromise. No thought or visual input is too small to have an effect.

Once we *identify* and then courageously *eliminate* the visual inputs that trigger our personal lust, we have eliminated, or at least greatly reduced the severity of, our struggles, even before the battle begins. Therefore, because I am a child of God, and because I reverence the sacredness of my mind, "*I will set no wicked thing before mine eyes*" (Ps. 101:3). I like to say: *I will allow no mental picture to hang on the walls of my imagination that I would not hang on the walls of my home.*

The pupil of the eye (the dark spot in the middle) was designed by God to expand to a full quarter of an inch to let in light when it's dark. But in bright light, it can protect the sensitive retina receptors by closing to the size of a pinhead. If God feels the delicate precision of the eye is that deserving of protection from damage by light, how deserving of protection is it from filthy pornography?

We've learned that no one can have the Spirit of the Lord while indulging eyes full of adultery or lust. The scriptures teach that the eternal consequences of misusing our bodies is so important that it would be better to crucify a disobedient body part than to indulge its bad habits and allow it to lead us into hell. (See Rom. 6:6; Gal. 5:24; 2:20.) Alma connected our need for visual discipline to the severity of the cross and the crucifixion when he counseled his adulterous son, "I would that ye should repent and forsake your sins, and *go no more after the lusts of your eyes*, but *cross yourself* in all these things; for except ye do this ye can in nowise inherit the Kingdom of God. Oh, remember, and take it upon you, and *cross yourself* in these things" (Alma 39:9).

Now consider the stark imagery used by Christ after condemning mental adultery: "And if thine eye offend thee, pluck it out, and cast it from thee: it is better for thee to enter into life with one eye, rather than having two eyes to be cast into hell fire" (Matt. 18:9; also Mark 9:47).

It's not likely that Jesus intended for people to actually destroy their vision by literally ripping their eyes from their sockets, although, as he said, even that drastic measure would be preferable to losing their exaltation. Surely these sobering words should teach us the importance of *controlling* our eyes (and the thoughts they stimulate) rather than allowing the lure of improper visual input to dictate what we view and think about. Jesus was most likely using this example to indicate the necessity for strict self-discipline in the way we use our eyes: the need for taking whatever action is required to control the lust of the eyes.

We can certainly apply the spirit of this lesson by "plucking away" the unacceptable input and thereby eliminating visual offenses to the Lord's Holy Spirit. For example, we're "plucking away" our eyes when we change the TV channel to avoid sexual provocation, when we walk out of a theater at the first sign of inappropriate entertainment, or turn our eyes away (as David should have) to prevent lusting after an attractive person. We're also "plucking away" visual offenses from our eyes (and reducing sexual struggles) when we refuse to view sensual magazines or Internet sites, movies, or videos that are incompatible with the Lord's Spirit. "Plucking away," then, means we must learn to prevent our eyes and our thoughts from lingering on anything that could fuel our appetites for inappropriate stimulation or imagination. As Job said, "I made a covenant with mine yes; why then should I *think* [fantasize] upon a maid?" (Job 31:1).

Of all the muscles in our bodies, the most active are those around the eyeball. In fact, it's been estimated that the eye muscles move an incredible 100,000 times every day! Perhaps we need to double that figure in order to develop spiritual control over our eyes as we learn to look away from things not appropriate.

EYES ASHAMED TO FACE GOD

Unrepented sins impose terrible guilt and shame upon us. This unnecessary dilemma damages our prayer life as we hang our heads in shame. "For innumerable evils have compassed me about," said one writer, "mine iniquities have taken hold upon me, *so that I am not able to look up*; they are more than the hairs of mine head: therefore my heart faileth me" (Ps. 40:12).

This shame that we feel *now* will be nothing compared to what we'll experience after a life of procrastination and indulgence. For if we come to that day of judgment unrepentant and unprepared, "in this awful state we shall *not dare to look up* to our God; and we would fain be glad if we could command the rocks and the mountains to fall upon us to hide us from his presence" (Alma 12:14). Sometimes we waste our probation and indulge in sinful foolishness because we believe Satan's deception that if we do it in secret, unknown to our fellow man, God won't know, and therefore it won't count. "Yet they say, The Lord shall not see, neither shall the God of Jacob regard it" (Ps.

94:7). "He hath said in his heart, God hath forgotten: he hideth his face; he will never see it" (Ps. 10:11).

The spiritual danger in thinking that our sins go unnoticed is that we begin to depend on that thought. We actually believe that we can "get away with it." "For thou hast trusted in thy wickedness: thou has said, None seeth me" (Isa. 47:10). The longer this attitude prevails, the worse our degeneracy becomes. "The iniquity of the house of Israel and Judah is exceeding great, and . . . the city full of perverseness: for they say . . . *the Lord seeth not*" (Ezek. 9:9).

Those who commit adultery and other sexual sins don't make public announcements of their sinful intent. Rather, they attempt to hide their acts in the darkness or in secret places where they assume they will be undiscovered. "The eye also of the adulterer waiteth for the twilight, saying, No eye shall see me: and disguiseth his face" (Job 24:15). But it's impossible to hide our sins from God, for "*the eyes of the Lord are in every place*, beholding the evil and the good" (Prov. 15:3). "Can any hide himself in secret places that I shall not see him? saith the Lord" (Jer. 23:24), "*for mine eyes are upon all their ways*: they are not hid from my face, neither is their iniquity hid from mine eyes" (Jer. 16:17). "Neither is there any creature that is not manifest in his sight: *but all things are naked and opened unto the eyes of him with whom we have to do*" (Heb. 4:13).

If we keep our eyes and thoughts clean and obedient, we can feel the same confidence Enos felt in meeting the Lord. "And I rejoice in the day when my mortal shall put on immortality, and shall stand before him; *then shall I see his face with pleasure*, and he will say unto me, Come unto me, ye blessed, there is a place prepared for you in the mansions of my Father" (Enos 1:27).

THE VEIL

Have you ever looked through sunglasses that made everything appear brown, green, bluish, or yellow? In the same way, things are not always as they appear when seen though mortal eyes. "For now we see through a glass, darkly; but then face to face: now I know in part; but then shall I know even as also I am known" (1 Cor. 13:12).

There are many spectrums of light, such as ultraviolet or radar waves, which our physical eyes cannot see or discern without the aid of specialized machines or instruments. And so it is with the spirit world, spirit matter, and spirit beings. "All spirit is matter, but it is more fine or pure, *and can only be discerned by purer eyes*" (D&C 131:7). Yet, even with the present limitations, our mortal eyes are capable of detecting ten million different colors! But the issue is: how many spiritual *principles* are they capable of seeing or discerning?

When we left the premortal world to come here, some kind of mental or optical filter was put in place to prevent us from seeing and remembering things that would distract or compromise our mortal probation. It's called *the veil*, "for there is no remembrance of former things" (Eccl. 1:11). While other realms are all around us, coexistent and sharing the same space, we don't see them through mortal eyes unless quickened by the Spirit to pierce that veil. For example, the spirit world where people go when their mortal bodies die is right here on this same planet, merged and coexistent in the same space that we occupy, but in a *spiritual* realm rather than physical.

"Where is the spirit world?" asked Brigham Young. "It is right here. . . . Do they go beyond boundaries of this organized earth? No, they do not. They are brought forth upon this earth, for the express purpose of inhabiting it to all eternity. Where else are you going? Nowhere else, only as you may be permitted" (*Journal of Discourses*, Vol. 3, p. 369, as quoted by Duane S. Crowther in *Life Everlasting* [Bookcraft, Salt Lake City: 1967], p.13). Since the spirits that inhabit the spirit world are all around us, Brigham Young asked on another occasion:

> Can you see spirits in this room? No. Suppose the Lord should touch your eyes that you might see, could you then see the spirits? Yes, as plainly as you now see bodies, as did the servant of [Elisha]. [See 2 Kings 6:16-17.] If the Lord would permit it, and it was his will that it should be done, you could see the spirits that have departed from this world, as plainly as you now see bodies with your natural eyes (*Discourses of Brigham Young*, pp. 376-377, as quoted in *Teachings of Presidents of the Church*, 1998-99 Priesthood Manual, p. 279).

Let's consider some other examples of the things that are all around us, but which remain unseen unless the veil is pierced. When the brother of Jared was petitioning the unembodied Christ to touch the stones and cause them to shine so that they might have light in their enclosed boats, the Lord stood before him, unseen. But as "the Lord stretched forth his hand and touched the stones one by one with his finger . . . *the veil was taken from off the eyes of the brother of Jared*, and he saw the finger of the Lord" (Ether 3:6). Piercing the veil to see things otherwise hidden is related to our righteousness and faith. "And because of the knowledge of this man he could not be kept from beholding within the veil" (Ether 3:19). But the brother of Jared was not the only one with such great faith, for "there were many whose faith was so exceedingly strong, even before Christ came, *who could not be kept from within the veil*, but truly saw with their eyes the things which they had beheld with an eye of faith, and they were glad" (Ether 12:19).

When young Enoch was called by God to be his prophet, he was filled with the normal apprehension that comes to all of us. Part of his tutoring on that occasion included an anointing of his eyes so that he could discern the vastness of the spiritual reality which normally remains hidden. "And he beheld the spirits that God had created; and he beheld also things which were not visible to the natural eye." (See Moses 6:31-36.)

President Joseph F. Smith experienced the same quickening as the veil was removed from his eyes. He, too, was allowed to see and understand the things of the spirit world. "As I pondered over these things which are written, *the eyes of my understanding were opened*, and the Spirit of the Lord rested upon me, and I saw the hosts of the dead, both small and great" (D&C 138:11). "And as I wondered, *my eyes were opened*, and my understanding quickened, and I perceived" (D&C 138:29).

Under similar circumstances, Abraham's mortal eyes were also quickened so that the veil was removed and he was able to see galaxies and distant creations that man would not discover for thousands of years until powerful telescopes were created to aid his limited vision. "And he [Jehovah] put his hand upon mine eyes, and I saw those things which his hands had made, which were many; and they multiplied before mine eyes, and I could not see the end thereof" (Abr. 3:12). Perhaps Moses best explained this divine quickening and removal of the veil. Unlike Abraham, whose quickened eyes saw *out-*

ward to the vastness of the universe, Moses saw *inward*, to view details of this earth: every inhabitant and even the atoms and molecular structure of the earth.

> And it came to pass, as the voice [of God] was still speaking, Moses cast his eyes and beheld the earth, yea, even all of it; a*nd there was not a particle of it which he did not behold*, discerning it by the spirit of God.
> And he beheld also the inhabitants thereof, and there was not a soul which he beheld not; and he discerned them by the Spirit of God; and their numbers were great, even numberless as the sand upon the sea shore (Moses 1:27-28).

Moses later explained *how* his mortal eyes were able to see the spiritual realm which God showed him through the veil:

> But now mine own eyes have beheld God; *but not my natural, but my spiritual eyes,* for my natural eyes could not have beheld; for I should have withered and died in his presence; but his glory was upon me; and I beheld his face, for I was transfigured before him (Moses 1:11).

In our time the Savior has advised us to "prepare for the revelation which is to come, *when the veil of the covering* of my temple, in my tabernacle, *which hideth the earth*, shall be taken off, and all flesh shall see me together" (D&C 101:23). Because we don't understand the nature of the veil, this language is difficult to understand. Additional clues are contained in the following revelation concerning the Second Coming: "And there shall be silence in heaven for the space of half an hour; and immediately after shall *the curtain of heaven be unfolded, as a scroll is unfolded after it is rolled up*, and the face of the Lord shall be unveiled" (D&C 88:95).

Our telescopes pierce millions of light years into space, so how there could be a curtain, or a veil, hiding the earth—or hiding those on the earth from other things to be seen—we don't understand. *How* the veil covers us from what we're not meant to see in our daily lives isn't explained in the revelations, but Paul gave a clue that it has something to do with the effect the mortal flesh has on the relationship between the connections of our *spirit* mind and our *physical* brain and eyes. He mentioned, in discussing other things, "through the veil, that is to say, his flesh" (Heb. 10:20). We can't help wishing he'd gone into detail. But knowing *how* the veil operates isn't as important as learning its effect upon us. The veil, which limits our memory and visual perceptions, allows us to focus upon our purpose and mission here. Imagine how distracting it would be to our mortal schooling if these spiritual realms were constantly open to our view! We might never learn to exercise the principle of faith.

But even though many things are hidden, God doesn't mean for all of them to *remain* that way and has asked us to seek greater vision. Paul taught how helpless we are to see the spiritual things we need to see and understand without the quickening influence of the Spirit upon our eyes and discernment.

> The God of our Lord Jesus Christ, the Father of glory, may give unto you the spirit of wisdom and revelation in the knowledge of him: The eyes of your understanding being enlightened; that ye may know what is the hope of his calling, and what the riches of the glory of his inheritance in the saints, And what is the exceeding greatness of his power to us-ward who believe, according to the working of his mighty power (Eph. 1:17-19).

The Lord has encouraged each of us to develop our faith and discernment so that we, too, may pierce the veil to see and understand things that will further our spiritual progress.

> And again, verily I say unto you that it is your privilege, and a promise I give unto you that have been ordained unto this ministry, that inasmuch as you strip yourselves from jealousies and fears, and humble yourselves before me, for ye are not sufficiently humble, *the veil shall be rent* and you shall see me and know that I am—not with the carnal neither natural mind, but with the spiritual (D&C 67:10).

But it's not only things "beyond the veil" that our eyes need quickening to see. Sometimes we need divine assistance to see things that are obvious and all around us, but which we don't see because we're preoccupied with other matters. For example, the Savior longs to encircle each of us in the arms of his love and convince us how important we are to him and our Heavenly Father. But he's often restrained from doing so by our discouragement and unbelief.

For example, the Prophet Joseph Smith once had a vision in which he saw nine of the twelve apostles in a foreign land. He saw them standing in a circle without shoes. They had been beaten. They were tattered and discouraged, looking at the ground in despair. Standing above them in the air was the Savior, reaching toward them, yearning to lift them, comfort, strengthen, and encourage them with the arms of his love. But they neither saw nor discerned his presence. Joseph said the Savior looked upon them and wept.

It's said that Joseph could never relate this vision without weeping himself. Speaking of this deep emotion, Truman G. Madsen asked, "Why? Why should he be so touched? Because Christ willingly came to the earth so that the Father's family could come to him boldly, knowing that he knows what is taking place in us when we sin, that he knows all our feelings and cares. The greatest tragedy of life," he said, "is that, having paid that awful price of suffering, 'according to the flesh that his bowels might be filled with compassion' (see Alma 7:11-13), and being now prepared to reach down and help us, he is forbidden because we won't let him. We look down instead of up" (*The Highest In Us* [Salt Lake City: Bookcraft, 1978], p. 85).

Part of our spiritual development involves an awareness that what we normally see is not the full reality that is there for us to see. We must learn to look beyond the obvious with a spiritual discernment that can show us greater and more important things than unenlightened eyes normally behold. One of the things we have difficulty seeing with mortal eyes is the divine purpose in adversity and the schooling circumstances which a loving God brings into our lives. Too often, they seem to us to be nothing but unwanted and discouraging problems. Asking us to trust his divine providence and good will as he watches over us and seeks to bless us, the Lord said, "*Ye cannot behold with your natural eyes*, for the present time the design of your God concerning those things which shall come hereafter, and the glory which shall follow after much tribulation" (D&C 58:3).

One of the ways we gain trust in God's dealings with *us* is by studying the scriptural accounts of his dealings with *others*. This study reveals who God is, the kind of person he is, and the principles by which he operates. Searching the scriptures is probably the fastest way "that the eyes of the people might be opened to see and know the goodness and glory of God" (Mosiah 27:22). "The commandment of the Lord is pure, enlightening the eyes" (Ps. 19:8).

Another way to build trust as we relate to the Lord through the veil is to remember that he has blessings and glories and joys awaiting us there which are beyond present comprehension. "But as it is written, Eye hath not seen, nor ear heard, neither have entered into the heart of man, the things which God hath prepared for them that love him" (1 Cor. 2:9; see also Isa. 64:4; 3 Ne. 17:16; D&C 133:45).

Understanding our need for divinely given spiritual discernment to see beyond the obvious, each time we read the scriptures, we might first pray with the psalmist: "Open thou mine eyes, that I may behold wondrous things out of thy law" (Ps. 119:18). "Let him that is ignorant learn wisdom by humbling himself and calling upon the Lord his God, *that his eyes may be opened that he may see,* and his ears opened that he may hear" (D&C 136:32).

SPIRITUAL BLINDNESS

> O how marvelous are the works of the Lord, and how long doth he suffer with his people; yea, and *how blind and impenetrable are the understandings of the children of men;* for they will not seek wisdom, neither do they desire that she should rule over them! (Mosiah 8:20).

In addition to the divinely imposed veil that filters the eyes and memory of every mortal being, there are two other veils that cause spiritual blindness. These two veils are not of divine origin. They are the *veil of darkness* and the *veil of unbelief.* (See Moses 7:61; D&C 38:8.) These two veils may come from such innocent origins as the lies or false traditions which we inherit from our parents and friends, or from succumbing to satanic attacks on our beliefs and attitudes. Or they can be self-imposed by our own choices and stupidity.

BLINDNESS FROM TRADITION

> These are they who are honorable men of the earth, who were blinded by the craftiness of men (D&C 76:75).

While visiting with our son's family, my wife fixed them a breakfast of pancakes. The children were astonished and protested that "no one" eats pancakes for breakfast. In their family tradition, pancakes were served for a simple meal on Sunday evenings. I'm not aware of any spiritual significance in when we eat our pancakes, but this experience illustrates how many things we believe merely because they were the traditions we received from our families or cultures.

> For there are many yet on the earth among all sects, parties, and denominations, *who are blinded by the subtle craftiness of men,* whereby they lie in wait to deceive, and who are only kept from the truth because they know not where to find it (D&C 123:12).

The Lamanites were deceived by the misconceptions and falsehoods inherited from the descendants of Laman and Lemuel. Because they were trained not to trust the descendants of Nephi, they simply could not believe their teachings. As King Lamoni was converted by the great missionary Ammon, the Book of Mormon reports that "*the dark veil of unbelief* was being cast away from his mind" (Alma 19:6).

> Behold, *when ye shall rend that veil of unbelief* which doth cause you to remain in your awful state of wickedness, and hardness of heart, and blindness of mind, then shall the great and marvelous things which have been hid up from the foundation of the world . . . (Ether 4:15).

BLINDNESS FROM SATAN

Lest we become free from the devil by comprehending and obeying truth, every time we try to improve our knowledge or practice of the gospel, Satan and his followers rush into our learning experience and do all within their power to discredit it or blind us to the truth. (See Matt. 13:19; Mark 4:14-15; Luke 8:11-12.) The "mists of darkness" that Lehi and Nephi saw in their visions of the Tree of Life "are the temptations of the devil, *which blindeth the eyes,* and hardeneth the hearts of the children of men, and leadeth them away into broad roads, that they perish and are lost" (1 Ne. 12:17). "Satan seeketh to turn their hearts away from the truth, *that they become blinded and understand not* the things which are prepared for them" (D&C 78:10; see also 3 Ne. 1:19-22; 2:2-3).

The Lord told Moses that after Lucifer, a great spiritual leader in the premortal life, lost the war in heaven and decided to become an enemy to God, "he became Satan, yea, even the devil, the father of all lies, [and dedicated himself] *to deceive and to blind men*, and to lead them captive at his will, even as many as would not hearken unto my voice" (Moses 4:4). "In whom the god of this world *hath blinded the minds of them which believe not*, lest the light of the glorious gospel of Christ, who is the image of God, should shine unto them" (2 Cor. 4:4).

Yielding to the temptations of Satan is like giving him permission to place a blindfold over our eyes. It gives him the power to darken our vision to the spiritual insights we need to achieve happiness and victory over our faults.

> O ye wicked and perverse generation, why hath Satan got such great hold upon your hearts? *Why will ye yield yourselves unto him that he may have power over you, to blind your eyes,* that ye will not understand the words which are spoken, according to their truth? (Alma 10:25).

The Book of Mormon records a later fulfillment of this very principle, when Mormon observes, "And thus did Satan get possession of the hearts of the people again *insomuch that he did blind their eyes* and lead them away to believe that the doctrine of Christ was a foolish and a vain thing" (3 Ne. 2:2).

SELF-IMPOSED BLINDNESS

Sometimes we know the truth, but we close our eyes. We don't want to see and come under obligation. When we deliberately close our eyes to hide from the truth, we impose spiritual darkness and confusion *on ourselves*. "We grope for the wall like the blind, and *we grope as if we had no eyes*: we stumble at noonday as in the night; we are in desolate places as dead men" (Isa. 59:10). "Son of man, thou dwellest in the midst of a rebellious house, which have eyes to see, and see not; they have ears to hear, and hear not: for they are a rebellious house" (Ezek. 12:2).

Amulek was such a man—a man who knew the truth and the ways of the Lord, but was inactive and disobedient until an angel changed his heart and prepared him to become a companion and assistant to Alma. His words are interesting. He said, "I have seen much of his mysteries and his marvelous power; yea, even in the preservation of the lives of this people. Nevertheless, I did harden my heart, for I was called many times and I would not hear; therefore *I knew concerning these things yet I would not know*; therefore I went on rebelling against God, in the wickedness of my heart" (Alma 10:5-6).

Amulek was like the people mourned by Christ, whose "heart is waxed gross, and their ears are dull of hearing, *and their eyes they have closed; lest at any time they should see with their eyes*, and hear with their ears and should understand with their heart, and should be converted, and I should heal them" (Matt. 13:15). "Wo unto the blind that will not see; for they shall perish also" (2 Ne. 9:32).

Choosing to close our eyes (so that we don't see the truths that would obligate us to change and improve) is defined by the Lord as sin: "But if he repent not of his sins, *which are* unbelief and blindness of heart . . . ." (D&C 58:15). Peter taught that when we ignore gospel principles and fail to develop the spiritual personalities and characteristics the gospel is meant to provide, we become "blind, and cannot see afar off" (2 Pet. 1:9). This self-imposed blindness truly angers the Lord because it frustrates the purpose of our coming here and wastes our probation. "For thus saith the Lord: I am angry with this people, and my fierce anger is kindled against them; for their hearts have waxed hard, and their ears are dull of hearing, and *their eyes cannot see afar off*" (Moses 6:27).

CHRIST RESTORES OUR SIGHT

Whatever the cause of our spiritual blindness and confusion, the Savior's role is to rescue us from darkness and lead us back into the light. While Satan is trying to blind our spiritual perception, Christ is working to protect and recover it. Jesus said, "For judgment I am come into this world, *that they which see not might see*" (John 9:39). As we come to the Lord in prayer, often confused and blind to the best alternatives, we may be confident in his desire to open our eyes to the right path and the best choices.

Jesus also said, "The Spirit of the Lord is upon me . . . he hath sent me to heal the brokenhearted, to preach deliverance to the captives, *and recovering of sight to the blind,* to set at liberty them that are bruised" (Luke 4:18). Isaiah prophesied that one purpose of the Savior's ministry would be "to open the blind eyes" (Isa. 42:7; also 35:5). "And I will bring the blind by a way that they knew not; I will lead them in paths that they have not known: I will make darkness light before them, and crooked things straight. These things will I do unto them, and not forsake them" (Isa. 42:16).

Prayer and scripture study are the keys to receiving Christ's help to improve our spiritual vision "that we might read and understand of his mysteries, and have his commandments always before our eyes" (Mosiah 1:5). The iron rod in Lehi's vision of the Tree of Life represents the word of God, or the scriptures. The message of that vision is that those who hold fast to the rod by staying close to the scriptures are protected from the temptations and mists of darkness that Satan tries to use to lead us into spiritual blindness. "And whoso would hearken unto the word of God, and would hold fast unto it, they would never perish; neither could the temptations and the fiery darts of the adversary *overpower them unto blindness,* to lead them away to destruction" (1 Ne. 15:24). "But blessed are your eyes, for they see" (Matt. 13:16).

EVERY EYE SHALL SEE

One of the most exciting events we have to look forward to is the Second Coming of Christ, when something will happen to people's eyes that has never happened before: "And prepare for the revelation which is to come, when the veil of the covering of my temple, in my tabernacle, which hideth the earth, shall be taken off, *and all flesh shall see me together*" (D&C 101:23).

"Behold," said John, "he cometh with clouds; *and every eye shall see him*" (Rev. 1:7). "There is none to escape," said Christ, for "*there is no eye that shall not see,* neither ear that shall not hear, neither heart that shall not be penetrated" (D&C 1:2). "And the glory of the Lord shall be revealed, and all flesh shall see it together: for the mouth of the Lord hath spoken it" (Isa. 40:5).

Whether Jesus will see his image in us at that time depends, in large measure, on what we're seeing with our eyes today. Whether we'll see him at that time of judgment with pleasure or dread also depends, to a great extent, on the use we make of our eyes today. "I say unto you, can ye look up to God at that day with a pure heart and clean hands?" (Alma 5:19).

Chapter Three

THE EARS

"Blessed are your . . . ears, for they hear" (Matt. 13:16).

It has been said that God gave us two ears and only one mouth so that we'd spend at least twice as much time *listening* as we do *talking*. I'm not sure that was the reason, but with more than two thousand revelations regarding the use of our ears, it's obvious that how we use our ears (or fail to use them) is of concern to the Lord. "Wherefore, my beloved brethren, let every man be *swift* to hear, *slow* to speak" (James 1:19).

The physical mechanisms for hearing sound are amazing. Sound waves are first gathered by the outside part of the ear and sent down the ear canal to vibrate the eardrum (tympanic membrane), a small membrane about the size of your little fingernail and thinner than your skin. These vibrations are then transferred to three tiny bones deep in the middle ear (known as the auditory ossicles), about the size of a grain of rice. These tiny levers amplify the sound vibrations by as much as twenty-two times before they enter the cochlea, a fluid-filled, coiled tube where tiny hairlike cells turn the amplified vibrations into nerve messages that are flashed (in about one hundredth of a second) to the brain for interpretation.

We can learn much about what the Lord expects us to do with *our* ears by observing what he does with *his*. Near the end of his ministry, the Savior was often accompanied by large crowds. "And there went great multitudes with him," Luke reported, even "an innumerable multitude of people insomuch that they trode upon one another" (Luke 14:25; 12:1).

We find a great lesson about ears on the Savior's way to his last Passover in Jerusalem. "And it came to pass, that as he was come nigh unto Jericho, a certain blind man sat by the way side begging: And hearing the multitude pass by, he asked what it meant" (Luke 18:35-36). In our society we're used to massive crowds and the accompanying noise, but such an occurrence would have been astonishing in the little town of Jericho.

It was the Savior's practice to teach as he traveled, so we can assume that he was occupied with such matters when the beggar called to him from the side of the road, saying, "Jesus, thou Son of David, have mercy on me" (Luke 18:38). Surrounded as he was by hundreds, if not thousands, of people, it's improbable that Christ could have even *seen* the beggar on the side of the street, much less *heard* him. But Christ's eyes and ears were always attuned to the needs of people. He did hear the blind man—even above the noise of the crowd. Jesus instantly stopped and "commanded him to be brought unto him: and when he was come near, he asked him, Saying, What wilt thou that I shall do unto thee? And he said, Lord, that I may receive my sight" (Luke 18:40-41). The request was granted, and the people surely marveled over the healing.

We should marvel no less at the power of Christ's ears to hear even the most timid request for help. Today a traveler can be preoccupied with urgent matters in an airport, paying no attention to the loudspeakers, and yet hear his name the instant he is paged. A mother can hear her baby's cry even when she is exhausted and sleeping. Just as Christ trained his ears to hear the important cry for help above the distracting noise of the crowd, we too can learn to hear the needs of those around us.

EARS THAT CAN HEAR

Many of us hear things *physically* even though we tune them out *mentally*. That is to say, we're conscious of the words being formed by the vibrations in our ears, but we're not allowing them to register in the I-need-to-do-something-about-this-message part of our brains because we're more interested in other thoughts. We're only half listening.

This inattention can lead to serious regrets. When the eleven brothers of Joseph, whom they sold into Egyptian slavery, thought back on his cries for mercy, they were plagued with great remorse because they had *heard* his cries but refused to hearken. "And they said one to another, We are verily guilty concerning our brother, in that *we saw the anguish of his soul, when he besought us, and we would not hear;* therefore is this distress come upon us" (Gen. 42:21).

When I worked in aerospace, a perpetually negative, unpopular, and rather offensive coworker spent the last few weeks of his life trying to tell us of his heartbreak at being divorced from a wife he still loved. His whining fell on uncaring ears. But when he committed suicide we had serious regrets, wishing we had been kinder and more caring; wondering if attentive, caring ears might have given him the strength to search for other alternatives.

The Lord needs our ears to be sensitive and caring in his service, to hear the cries of his children for help, both spoken and unspoken. You and I can prevent such remorse by making sure that no one in our family or circle of influence will ever accuse us of not hearing or recognizing their cries for help.

Sometimes the kindest thing we can do for another person is to drop what we're doing and give the speaker our full, immediate attention. I learned how riveting this can be when I went to see a real estate broker about the poor service I felt we were receiving from his salespeople. When I arrived for the appointment, I was astonished by his actions. I expected a battle. I expected him to defend his people's poor results. I expected him to try to persuade me to be more patient. Instead, he closed the door to his office, called his secretary, and told her that he was not to be disturbed. Then he unplugged his phone, smiled, and said: "Now, how can I help you?" There was no question that he was giving me full attention and that for the next period of time, "his ears were all mine."

A pilot flew his single-engine plane to a small community where the only place to land was a dirt strip cleared in a field. He arrived over the field just as the sun was setting behind a mountain. By the time he'd circled for a landing, it was too dark to be sure of the position of the clearing. Of course there were no lights to mark the landing strip, nor was there a control tower or radio to guide him in. To make matters even worse, the plane's landing lights didn't work, so there was little chance that the pilot could land without crashing. He was completely helpless. This poor man circled that field for two hours, not knowing what else to do, expecting tragedy as soon as he ran out of fuel.

But then a miracle occurred. Someone on the ground used his ears to detect his dilemma. Noticing the continuing drone of the plane's engine, a perceptive man realized the pilot's predicament. He jumped in his car, hurried to the field, and drove up and down the landing strip several times to indicate its boundaries with his headlights. Then he parked at one end so that his car's lights could mark the beginning of the landing strip. Thanks to this man hearing his need, the pilot made a safe landing. (Adapted from a Focus on the Family Newsletter; Arcadia, California: Aug. 1985, p. 16.) What an interesting and unusual rescue!

Many people we know are in painful or frightening "holding patterns" that they can't communicate to us. When God needs us to hear their clues and their unspecific cries for help, will we have the ability to discern and respond to their need?

GOD HEARS EVERY PRAYER

"But when ye pray, use not vain repetitions, as the heathen, for *they think that they shall be heard for their much speaking*" (3 Ne. 13:7; also Matt. 6:7).

We don't need any special "techniques" or manipulations to persuade God to hear our prayers. One thing we know about Heavenly Father is that he is ever watchful over our affairs, looking and listening attentively for opportunities to guide, inspire, protect and comfort us in our daily struggles. "The eyes of the Lord are upon the righteous, and *his ears are open unto their cry*" (Ps. 34:15). This is what he does, not because we manipulate him, but simply because we, as his children in mortal school, are among his highest priorities. He loves us. His life's work is to observe, listen, answer, and guide us. (See Moses 1:39.) "The righteous cry, and the Lord heareth, and delivereth them out of all their troubles" (Ps. 34:17).

Repeated testimonies throughout the ages bear witness to the fact that God hears and answers prayers. For example, an angel who was sent in response to Daniel's prayer said: "Fear not, Daniel: for *from the first day* that thou didst set thine heart to understand, and to chasten thyself before thy God, *thy words were heard*, and I am come for thy words" (Dan. 10:12). Similarly, the Lord said to Solomon: "I have heard thy prayer and thy supplication, that thou hast made before me" (1 Kgs. 9:3). When the angel appeared to Zacharias in the temple, his fear was calmed with these words: "Fear not, Zacharias: *for thy prayer is heard"* (Luke 1:13). Cornelius, a Roman centurion who became the first gentile convert to Christianity, was also told by an angel, *"Cornelius, thy prayer is heard*, and thine alms are had in remembrance in the sight of God" (Acts 10:31). "But verily God hath heard me; he hath attended to the voice of my prayer" (Ps. 66:19).

When an angel was sent to teach King Benjamin, his message was prefaced with these words: *"For the Lord hath heard thy prayers*, and hath judged of thy righteousness, and sent me to declare unto thee that thou mayest rejoice" (Mosiah 3:4). The angel sent to rescue the Prophet's apostate son, Alma, explained why he was there: "Behold, *the Lord hath heard the prayers of his people*, and also the prayers of his servant, Alma, who is thy father; for he has prayed with much faith concerning thee" (Mosiah 27:14).

In our own time, the Lord said to Joseph Smith, "Thy prayers and the prayers of thy brethren *have come up into my ears"* (D&C 90:1). To Sidney Gilbert he said, "I have heard your prayers" (D&C 53:1). And to Thomas B Marsh: "I have heard thy prayers; and thine alms have come up as a memorial before me . . . I know thy heart, and have heard thy prayers concerning thy brethren" (D&C 112:1, 11). "Rejoice evermore, and in everything give thanks: Waiting patiently on the Lord, for *your prayers have entered into the ears of the Lord* of Sabaoth, and are recorded with this seal and testament—the Lord hath sworn and decreed that they shall be granted" (D&C 98:1-2).

One of the saddest conditions of our modern society is the abuse inflicted on wives and children by cruel husbands and fathers. Their day of accountability will surely come, but meanwhile, the wounds and cries of these victims do not go unnoticed by the Lord. "I, the Lord, *have seen the sorrow, and heard the mourning of the daughters of my people. . .* because of the wickedness and abominations of their husbands" (Jacob 2:31).

We've been given many instructions for the care of the poor and needy so "that the cries of the widow and the fatherless come not up into the ears of the Lord against this people" (D&C 136:8). It's comforting to know that his attention is particularly directed toward the most needy. The Lord specifically stated that if "any widow, or fatherless child. . . cry at all unto me, *I will surely hear their cry*" (Ex. 22:22-23). Where is *our* attention directed?

WHY ANSWERS ARE SOMETIMES DELAYED

Many of us struggle to gain our own testimonies that God hears our prayers. Even though the Lord is *"quick to hear* the cries of his people" (Alma 9:26), the answers come in the Lord's time, which is seldom as quick as we wish for. For example, the very first message to Moses as he encountered Jehovah at the burning bush was a message concerning the Lord's ears:

> And the Lord said, *I have surely seen the affliction of my people* which are in Egypt, *and have heard their cry* by reason of their taskmasters, *for I know their sorrows*; And I am come down to deliver them out of the hand of the Egyptians. . . . Now therefore, behold, *the cry of the children of Israel is come unto me* (Ex. 3:7-9).

Yet, for various reasons, those groanings and prayers spanned approximately four hundred years before deliverance was granted. Let's discuss three reasons why God hears, but may not send immediate answers: our *unworthiness,* our *lack of preparation*, and *our need to have our faith tested* to help us grow.

BECAUSE WE ARE UNWORTHY

"If I regard iniquity in my heart, the Lord will not hear me" (Ps. 66:18).

Our natural-man tendency is to turn away from people who repeatedly betray us. We invent stories and sayings to justify this impatience, like "The Boy Who Cried Wolf" and "The things you do shout so loudly in my ears that I can't hear what you say." But God never turns his ears from us. They are always attuned to our needs, waiting patiently to hear the tiniest clue that we might be ready to soften our hearts and try again.

> But after they had rest, they did evil again before thee: therefore leftest thou them in the hand of their enemies, so that they had the dominion over them: *yet when they returned*, and cried unto thee, *thou heardest them from heaven;* and many times didst thou deliver them according to thy mercies (Neh. 9:28).

While there's absolutely nothing we can do to diminish God's perfect and unconditional love for us, *our deliberate sin and rebellion can affect his hearing.* Abinadi was sent to warn King Noah and his people of their wickedness. "Go forth, and say unto this people, thus saith the Lord—Wo unto this people, for I have seen their abominations, and their wickedness, and their whoredoms; and except they repent I will visit them in mine anger" (Mosiah 11:20). And then the Lord explained what deliberate, unrepented rebellion can do to his hearing:

> Yea, and it shall come to pass that when they shall cry unto me *I will be slow to hear their cries;* yea, and I will suffer them that they be smitten by their enemies. And except they repent in sackcloth and ashes, and cry mightily to the Lord their God, *I will not hear their prayers, neither will I deliver them out of their afflictions* (Mosiah 11:24- 25).

His love is unceasing, but when we deliberately make wrong choices, he lovingly withdraws his blessings so that the natural consequences of our poor choices can tutor us back to his ways. When Isaiah's wicked and unrepentant people questioned why the Lord did not respond to their prayers, the prophet explained the important principle we should remember: "Behold, the Lord's hand is not shortened, that it cannot save; neither his ear heavy, that it cannot hear: *But your iniquities have separated between you and your God,* and your sins have hid his face from you, that he will not hear" (Isa. 59:1). "I the Lord am bound when ye do what I say; but when ye do not what I say, ye have no promise" (D&C 82:10).

BECAUSE WE ARE NOT PREPARED

Our answers are often delayed, even when our requests are righteous and appropriate, because we're spiritually unprepared and would be harmed by premature answers. In such cases, God's divine love and foreknowledge mandates that some prerequisites be resolved in our lives before he responds to those desires.

> I don't believe he ignores his children when they talk to him. The problem in our communication with him is that not all of us have learned how to listen for his answers, *or perhaps we are not prepared to hear him.* I believe we receive his answers as we prepare ourselves to receive them (M. Russell Ballard, *Ensign,* June 1983, p. 73).

Delays in receiving the answers to our prayers can be less frustrating if we remember the Lord's promise: "I will order all things for your good *as fast as ye are able to receive them"* (D&C 111:11). As Elder Boyd K. Packer explained:

> Sometimes you may struggle with a problem and not get an answer. What could be wrong? It may be that you are not doing anything wrong. *It may be that you have not done the right things long enough.* Remember, you cannot force spiritual things. Put difficult questions in the back of your minds and go about your lives. Ponder and pray quietly and persistently about them. The answer may come as an inspiration, here a little and there a little, "line upon line and precept upon precept" (D&C 98:12). *(Ensign,* Nov. 1979, p. 21).

It's difficult for us to perceive the Father's eternal perspective and reasons for delaying our answers when we're caught up in the urgency of our immediate circumstances. H. Burke Peterson, of the presiding Bishopric, taught us to pray for understanding of *why* answers are delayed.

> May I suggest that when you pray for something very special, you pray for two things. First, pray for the blessing that you want, whether it's a new baby, or a job, or whatever; and second, ask the Lord for the blessing of understanding. Then, *if he feels for some reason that the blessing isn't appropriate for that time, the blessing of understanding will come*—and the frustrations that ofttimes come because we feel our prayers are not answered will blow away in the wind *(Ensign,* June 1981, pp. 74-75; emphasis added).

TO TEST OUR FAITH

When we pray worthily, we can count on receiving one of three answers. It may be a "Yes" or it may be a "No." But in many cases, the answer is "Not yet." Mortality is a probation. We're being tried and tested, and "ye receive no witness until after the trial of your faith" (Ether 12:6). Perhaps a delayed answer can be the Lord's way of saying, "Let's wait awhile and see if you will still trust me." Elder James E. Talmage explained that "in mercy the Father sometimes delays the granting that the asking may be more fervent" (*Jesus The Christ* [Salt Lake City: Deseret Book, 1956], p. 435). In other words, when the answers to our prayers are delayed, it's never because the Lord hasn't *heard*, but because he has wise and divine reasons for having us wait. It may be his way of saying, "Why don't you show me how much you care, so that I can show you how much I care?" He knows that the more we plead, the sweeter the joy will be when the answer does come. Sister Patricia Holland said:

> I want you to know that in my life, when I have had disappointments and delays, I have lived to see that if I continue to knock with unshakable faith and persist in my patience—waiting upon the Lord and his calendar— I have discovered that the Lord's "no's" are merely preludes to an even greater "yes" (*Speeches of the Year* [Provo, Utah: BYU Press, 1988-89], p. 73).

ASK GOD TO HEAR YOUR PRAYERS

"Lord, hear my voice: let thine ears be attentive to the voice of my supplications" (Ps. 130:2).

Asking Heavenly Father to hear our prayers is not an expression of doubt, but of faith. At the dedication of the Kirtland Temple, the Prophet Joseph Smith implored: "O Lord God Almighty, hear us in these our petitions, and answer us from heaven . . . O hear, O hear, O hear us, O Lord! And answer these petitions. . . ." (D&C 109:77-78). Similarly, at the dedication of Israel's first temple, Solomon's request on behalf of his people was: "If they return to thee with all their heart and with all their soul . . . Then hear thou from the heavens, even from thy dwelling place, their prayer and their supplications, and maintain their cause, and forgive thy people which have sinned against thee" (2 Chr. 6:38-39).

If it's appropriate to ask God to hear our prayers at the dedication of temple buildings, surely it's appropriate to do the same as we dedicate and rededicate our temple bodies to a higher level of spiritual commitment. "Hear my prayer, O Lord, and give ear unto my cry; hold not thy peace at my tears" (Ps. 39:12). Nehemiah prayed: "Let thine ear now be attentive, and thine eyes open, that thou mayest hear the prayer of thy servant" (Neh. 1:6). It would be difficult to imagine that any prayer offered from within the walls of a temple could be ignored by the Lord. And, as we progressively make temples of our bodies, the prayers uttered therein will become equally powerful and compelling. "O my God, incline thine ear, and hear; open thine eyes, and behold our desolations . . . *for we do not present our supplications before thee for our righteousness, but for thy great mercies*" (Dan. 9:18).

Sometimes the urgency of a situation won't allow time to wait for answers. For example, when Abraham's enemies grabbed him, tied him to an altar, and prepared to kill him in sacrifice to their false gods, there was not time to wait patiently for a "someday" answer to his cry for help. "And as they lifted up their hands upon me, that they might offer me up and take away my life," said Abraham, "behold, I lifted up my voice unto the Lord my God, *and the Lord hearkened and heard*, and he filled me with the vision of the Almighty, and the angel of his presence stood by me, and immediately unloosed my bands; And his voice was unto me: Abraham, Abraham, behold, my name is Jehovah, and I have heard thee, and have come down to deliver thee" (Abr. 1:15-16). What a comfort this must have been, not just to have his life saved, but to receive such immediate proof of God's personal love and attention to his needs! While we're admonished to be patient in our prayers, scriptures do justify requesting immediate attention when circumstances are urgent. "And hide not thy face from thy servant; *for I am in trouble: hear me speedily*" (Ps. 69:17).

DANGERS TO EARS

The things we allow our ears to hear are important because of their power to change our emotions, attitudes, values, and priorities. Words can nourish and strengthen us, or they can lead to spiritual defeat. As we seek to increase the sensitivity of our ears, we must be aware that we're surrounded by "conspiring men" as well as spiritual enemies who would put harmful suggestions into our minds and hearts through our ears.

All our lives we've heard our mothers warn us not to poke things in our ears. I once examined a man with hearing problems and found that his eardrum had been ruptured. He told me an interesting story. It seems his ear over-produced ear wax, which he had to clean frequently or it would accumulate and plug his ear. To dissolve the wax, he used a solution of hydrogen peroxide and one of those large cotton swabs about 8-10 inches long. One day as he was soaking his ear, his wife insisted that they leave on an errand. He left the swab soaking in the ear, planning to finish the job in the car. As he shut the door to the house, he was too close and jammed the swab through the eardrum.

It would be easy to condemn him for his carelessness, but let's consider the things we allow into our ears, such as dirty jokes, inappropriate language, sensational talk shows, or music that grieves the Holy Spirit and drives it from us. Such things can cause spiritual damage far worse than a ruptured eardrum.

Spirit beings, both demonic and divine, can speak to the spirits within mortals and cause thoughts to occur in their conscious minds, even though those individuals are not consciously aware of the communication or the source of the ideas. The spirit beings do this by "whispering in our ears." This is the way inspiration is received from the Holy Ghost—*and* the way temptations are presented by Satan and his demons. (See 2 Ne. 28:20-22.) As Sister Ardeth G. Kapp said, "The Holy Ghost communicates with us through 'whisperings' just as Satan and other spirit beings do" (*Ensign*, Nov. 1990, p. 94). Mormon related the influence of the Spirit's "whisperings" in guiding him as he condensed the sacred records of centuries into the Book of Mormon we have today: "I do this for a wise purpose; *for thus it whispereth me*, according to the workings of the Spirit of the Lord which is in me" (W of M 1:7).

Unless we're constantly on guard, Satan and his demons can whisper damaging and discouraging things to our spirits twenty-four hours a day without our even being aware of it. "And behold, others he flattereth away and telleth them there is no hell; and he saith unto them: I am no devil, for there is none—*and thus he whispereth in their ears*, until he grasps them with his awful chains" (2 Ne. 28:22).

Satan is a master at manipulating mortal thoughts and emotions through the vicious whisperings he has perfected with practice on billions of our predecessors. He knows how to stimulate worry and doubt, fear and hatred, resentment and jealousy, stress, anger, lust, depression, feelings of emptiness and loneliness, the desire to give up and retreat from life's challenges, and other unwholesome and discouraging emotions. No such feelings come from above. They come from the influence of powerful demonic spirits who know how to whisper suggestions that trigger our self-defeating emotions and thought patterns and then laugh and rejoice at our misery.

> Thoughts originate from three sources—from within us, from the prompting of the Holy Spirit, and from the evil sources around us provided by Satan and his hosts as they "whisper in our ears." We must, therefore, learn to recognize the source of our thoughts and control them accordingly (Ronald A. Dalley, *The New Era*, Aug. 1984, p. 44).

We now understand that all messages from the other side of the veil, whether they come from Satan and his army or from the Holy Ghost and other righteous messengers, are communicated directly to our spirits and then relayed into our conscious minds as thoughts, impressions, and suggestions. These whisperings then appear in our minds as if they were our own thoughts. Joseph Fielding Smith, Jr. warned of the danger this can pose:

> We should be on guard always to resist Satan's advances. He has power to place thoughts in our minds and to whisper to us in unspoken impressions to entice us to satisfy our appetites or desires and in various other ways he plays upon our weaknesses and desires *(Answers to Gospel Questions,* comp. 5 vols. [Salt Lake City: Deseret Book Co., 1957-66], vol. 3:81).

While the ability of Satan's army to surround us and whisper lies and temptations is alarming, we should rejoice in the knowledge that Satan cannot force us to listen to his lies. He can have absolutely no power over our minds and hearts except as we allow. And through the influence of the Holy Ghost, we can discern and reject all of Satan's evil suggestions. As Lawrence R. Peterson, Jr. explained:

As a being of spirit, he works in the realm of spirit, counterbalanced by the Spirit of God. In this way, free agency is preserved, giving us a choice between good and evil. As Lehi taught, "Man could not act for himself save it should be that he was enticed by the one or the other" (2 Ne. 2:16).

If Satan entices us to do evil so the Holy Spirit "entices" us to virtue. (See Mosiah 3:19). *Free agency demands that neither the Holy Spirit nor the evil spirit have power to control the person against his will (Ensign, July 1984, p. 31).*

Learning to ignore Satan's whisperings so that we can listen to the Spirit's is much like tuning a radio. The air is filled with radio waves broadcasting every conceivable type of program. Since a radio can play only one station at a time, however, we must decide which program to tune into. And so it is with our ears: either we tune into the Lord's frequency or Satan's. Both sides are "broadcasting" urgent messages to influence our thoughts, feelings, and choices. Neither side can compel us to listen to the messages they send. We have both the agency and the responsibility to choose which signals we allow our minds to focus on. As Boyd K. Packer stated, this we *can* do and this we *must* do:

All inspiration does not come from God. (See D&C 46:7.) The evil one has the power to tap into those channels of revelation and send conflicting signals which can mislead and confuse us. There are promptings from evil sources which are so carefully counterfeited as to deceive even the very elect. (See Matt. 24:4.) *Nevertheless, we can learn to discern these spirits (Ensign, Nov. 1989, p. 14).*

I will tell you in your mind and in your heart, by the Holy Ghost, which shall come upon you and which shall dwell in your heart. Now, behold, this is the spirit of revelation (D&C 8:2-3).

SPIRITUAL DEAFNESS

"And wo unto the deaf that will not hear; for they shall perish" (2 Ne. 9:31).

Have you ever seen people put their hands over their ears so they won't hear something? Many of us do that spiritually. While there are millions of people whose defective ears prevent them from hearing sounds, there are even more people who deliberately *choose* to be deaf—spiritually. "But they refused to hearken . . . and stopped their ears, that they should not hear" (Zech. 7:11). "But they obeyed not, neither inclined their ear, but made their neck stiff, *that they might not hear, nor receive instruction*" (Jer. 17:23).

This deliberate shutdown of the auditory system is an attempt to avoid the obligation of obedience that results when we allow the word of the Lord to be heard and take root inside us. Amulek was such a man. He knew the truth and the ways of the Lord, but was resistant and disobedient until an angel changed his heart and prepared him to become a companion and assistant to Alma. His words are interesting. "I have seen much of his mysteries and his marvelous power," he said, "Nevertheless, I did harden my heart, for *I was called many times and I would not hear;* therefore, *I knew concerning these things yet I would not know;* therefore I went on rebelling against God, in the wickedness of my heart" (Alma 10:5-6). "For this people's heart is waxed gross, and their ears are dull of hearing, and their eyes have they closed; *lest at any time* they should see with their eyes, and hear with their ears, and *should understand with their heart, and should be converted, and I should heal them*" (Matt. 13:15).

Today it's common for missionaries to have doors slammed in their faces by people who refuse to hear. It has always been so. "Ye stiffnecked and uncircumcised in heart and ears, ye do always resist the Holy Ghost: as your fathers did, so do ye" (Acts 7:51). "And it came to pass as he began to expound these things unto them *they were angry with him, and began to mock him; and they would not hear* the words which he spake" (Alma 21:10).

Jeremiah said that people "which have ears, and hear not" are "foolish people, and without understanding" (Jer. 5:21). This foolishness will lead them into spiritual death unless they repent, for they "that will not hear my voice but harden their hearts . . . wo, wo, wo, is their doom" (D&C 38:6).

Two consequences of deliberate spiritual deafness are the voiding of the atonement in our behalf and the ineffectiveness of our personal prayers, "for behold, my blood shall not cleanse them if they hear me not" (D&C 29:17) and "He that turneth away his ear from hearing the law, even his prayer shall be abomination" (Prov. 28:9). "And the arm of the Lord shall be revealed; *and the day cometh that they who will not hear the voice of the Lord,* neither the voice of his servants, neither give heed to the words of the prophets and apostles, *shall be cut off from among the people*" (D&C 1:14).

When King Lamoni was converted by Ammon, he began "to teach them the words which he had heard from the mouth of Ammon; *and as many as heard his words believed,* and were converted unto the Lord. But there were many among them *who would not hear his words; therefore they went their way*" (Alma 19:31-32). It was the same in the Old Testament: "Yet they obeyed not, nor inclined their ear, *but walked every one in the imagination of their evil heart*" (Jer. 11:8). This is the issue: Whether we will hear the word of the Lord and walk in *his* paths, or refuse to hear and walk in the paths we choose for ourselves. "But they hearkened not, nor inclined their ear, but walked in the counsels and in the imagination of their evil heart, *and went backward, and not forward*" (Jer. 7:24).

One way we can "go backward" is by listening to the wrong messages. Many philosophies, doctrines, voices, and ideas compete for the attention of our ears in this day of "conspiring men." Jesus said that those of us who are too busy or preoccupied with other matters to hear his words are not really committed to his kingdom, for "mine elect *hear my voice* and harden not their hearts" (D&C 29:7). But "whoso *receiveth not my voice* is not acquainted with my voice, and *is not of me*" (D&C 84:52). "He that is of God heareth God's words: ye therefore hear them not, because ye are not of God" (John 8:47).

Another way we can go backward in our spiritual quest is by closing our ears to suggestions about improving our discipleship. It's easy to take offense when we hear someone correcting us. It's the "natural-man" thing to do. It's as if our ears have a filtering device to shut out any opinion that would require us to change and grow. "He that refuseth instruction despiseth his own soul: but he that heareth reproof getteth understanding" (Prov. 15:32). Alma described this character trait as being "easy to be entreated" (Alma 7:23). "The ear that heareth the reproof of life abideth among the wise" (Prov. 15:31).

Old Testament prophets foresaw many of the miracles that Christ would perform, including the healing touch that would cause "the ears of the deaf [to] be unstopped" (Isa. 35:5). As his healing miracles were observed, the people "were beyond measure astonished, saying, He hath done all things well: he maketh both the deaf to hear, and the dumb to speak" (Mark 7:37). Certainly this was fulfilled by the *physical* healing performed on defective ears, but it also occurred *spiritually*, as hardened hearts were softened and closed minds were opened, so that people who were spiritually blind and deaf could finally see and hear the truth.

In our time, the priesthood brethren are expected to carry forth the Lord's healing work. The Lord said, "In my name they shall open the eyes of the blind, and unstop the ears of the deaf" (D&C 35:9). Ears that are stopped, or spiritually deaf to the Lord's message, can sometimes be opened by concerned disciples who find ways past spiritual barriers by speaking patiently, over time, with love, compassion, and the power of the Holy Spirit. "In my name they shall open the eyes of the blind, and unstop the ears of the deaf" (D&C 84:69).

LISTENING EARS

The Savior said that those disciples who "shall abide the day of my coming" will be those whose ears "hear my voice . . . and shall not be asleep" (D&C 35:21). His counsel, then, to those who want to spend eternity with him is to listen to him *now*. There is urgency to hearing him *now*, so that we don't lose growing and developing time by waiting to listen and respond *later*. Therefore, the Lord pleads, "hear my voice while it is called *today*" (D&C 45:6).

The Doctrine and Covenants is full of exhortations to listen. For example: "Learn of me, and *listen* to my words" (D&C 19:23). "*Listen* to the counsel which I shall give unto you" (D&C 100:2). "*Listen* to him who is the advocate with the Father, who is pleading your cause before him" (D&C 45:3). "*Listen* to the voice of Jesus Christ, your Lord, your God, and your Redeemer" (D&C 27:1). This constant reiteration of the principle of listening to the Lord indicates how important it is.

Wouldn't it be wonderful if we could get more faith by simply flipping a switch, taking a pill, or waving a magic wand? Faith can't be increased in those ways, but there are other ways we can increase our faith, and one of them involves our ears: "So then faith cometh by hearing, and hearing by the word of God" (Rom. 10:17).

This scripture is expressing a mathematical formula: the *more* we read, study, and hear the word of God, the more our faith will increase, while the *less* time and attention we give to hearing the word, the less our faith will grow. In fact, it will diminish, for "how shall they believe in him of whom they have not heard?" (Rom. 10:14). The mission of Christ is to "save all men if they will hearken unto his voice" (2 Ne. 9:21), but how can they hearken or respond if they don't first hear him? Heavenly Father's counsel is the same: *Listen to Christ.*

> And there came a voice out of the cloud, saying, This is my beloved Son: *hear* him (Luke 9:35; also Matt. 17:5).

> Behold my Beloved Son, in whom I am well pleased in whom I have glorified my name—hear ye *him* (3 Ne. 11:7).

> One of them spake unto me, calling me by name and said, pointing to the other—This is My Beloved Son. Hear *Him!* (JS-H 1:17).

We see the relationship between hearing and faith in the experience of Enos, who hungered for a closer relationship with the Lord. His faith was weak, being based on nothing but the words he'd heard from his father. But the faith he gained by hearing second-hand was enough to move him to action. After long and earnest prayer, he did hear the word of the Lord to him personally. He not only rejoiced in this revelation but reported the expansion of his faith: "And after I, Enos, had heard these words, *my faith began to be unshaken in the Lord*" (Enos 1:11). Our faith can also become unshaken when we learn to use our ears more attentively to hear the word of the Lord at every opportunity.

Jacob yearned to help the wicked people of his time repent. But he knew if that were ever to happen, it would require the participation of their ears: "O that ye would *listen* unto the word of his commands, and let not this pride of your hearts [that closes the ears] destroy your souls!" (Jacob 2:16). That's exactly what happened to Cain, who was offended and withdrew from fellowship with God's people. He behaved as many of us do when we're offended and close our ears to the very words we need the most: "And Cain was wroth, and *listened not* any more to the voice of the Lord" (Moses 5:26).

If we consider ourselves disciples of the Savior, our duty is (and our desire should be) to listen, hearken, yearn for, seek to hear, learn, and then do his will in each day's events. "Every one that is of the truth heareth my voice" (John 18:37). "Hearken, O ye people, and open your hearts and give ear from afar; and *listen, you that call yourselves the people of the Lord,* and hear the word of the Lord and his will concerning you" (D&C 63:1).

Hearing God's word is the starting point, but audio reception is not enough. We must then implant the words heard into the fiber of our minds so that action results and growth is achieved. Thus Jesus emphasized, "Let these sayings *sink down* into your ears" (Luke 9:44) and "blessed are they that hear the word of God, *and keep it*" (Luke 11:28). Keeping the word of God means "being not a forgetful hearer" (James 1:25) as well as obedience. "*Remember* therefore how thou hast received and heard, and *hold fast* and repent" (Rev. 3:3). "For not the *hearers* of the law are just before God, the *doers* of the law shall be justified" (Rom. 2:13). "Therefore we ought to *give the more earnest heed* to the things which we have heard, *lest at any time we should let them slip*" (Heb. 2:1).

Periodically, when our cars get a little sluggish, having lost power or acceleration, we take them to the mechanic to be tuned up so that the performance standards improve. Perhaps we should consider a spiritual tuneup for our ears. Just as we'd go to a professional for a hearing aid if our ears were physically deficient, we've been encouraged to come to the Lord for divine assistance to open our ears and quicken our hearing ability.

"Let him that is ignorant learn wisdom by humbling himself and calling upon the Lord his God, that his eyes may be opened that he may see, *and his ears opened that he may hear*" (D&C 136:32). This scripture is phrased not as a suggestion, but as a commandment. I take this to mean that we should actually be using such words in our prayers. We should be specifically asking for a quickening of our ears so that as we study scriptures and listen to gospel sermons and classes, we may hear and learn beyond the obvious vibrations that normal, untuned ears can hear. "I have not commanded you to come up hither to trifle with the words which I shall speak, but that you should hearken unto me, and *open your ears* that ye may hear, and your hearts that ye may understand, and your minds that the mysteries of God may be unfolded to your view" (Mosiah 2:9). "The Lord God hath opened mine ear, and I was not rebellious, neither turned away back" (2 Ne. 7:5).

Chapter Four

THE MOUTH

"But the tongue . . . is an unruly evil, full of deadly poison" (James 3:8).

In Brunswick, Georgia, a thief who stole a woman's purse was quickly apprehended and taken back to the shopping center so the victim could identify him. On the way, the detective explained to the thief that when they arrived, he was to get out of the police car and face the woman so they could get a positive ID. Detective Chris Stewart reported that "the suspect did exactly as he had been told. He stepped from the car and looked up at the victim. And before anyone could say anything, he blurted out, 'Yeah, that's her . . . that's the woman I robbed'" (*America's Dumbest Criminals* [Nashville, Tennessee: Rutledge Hill Press, 1995], pp. 19-20). "For by thy words thou shalt be justified, and by thy words thou shalt be condemned" (Matt. 12:37).

When King Benjamin said, "I cannot tell you all the things whereby ye may commit sin; for there are divers ways and means, even so many that I cannot number them," he also warned us about our words. "But this much I can tell you," he said, "that *if you do not watch yourselves*, and your thoughts, *and your words* . . . ye must perish. And now, O man, remember, and perish not" (Mosiah 4:29-30).

The scriptures detail many of the wasteful ways that we displease the Lord with our mouths and for which he will hold us accountable: "light speeches" (D&C 88:121); "foolish talking" (Eph. 5:4); "profane and vain babblings" (1 Tim. 6:20); "vain talking" (Titus 1:10);

"unprofitable talking" (Job 15:3); "great swelling words of vanity" (2 Pet. 2:18); "lying words" (Jer. 7:8); "beguiling and enticing words" (Col. 2:4); and "deceiving and vain words" (Eph. 5:6).

Those who waste their words carelessly, or use them as weapons to injure others, should ponder the fact that no one can hide the faults of their tongues, for "there is not a word in my tongue, but, lo, O Lord, thou knowest it altogether" (Ps. 139:4). One major thing we will all be held accountable for and judged by at the last day will be our words; not just our chosen, planned words, but every foolish and idle word. As Jesus warned, "I say unto you, That every idle word that men shall speak, they shall give account thereof in the day of judgment" (Matt. 12:36).

> Dearest children, holy angels
> Watch your actions night and day,
> And they keep a faithful record
> Of the good and bad you say.
> (Hymns of The Church of Jesus Christ of Latter-day Saints [Salt Lake
> City: The Church of Jesus Christ of Latter-day Saints, 1985], No. 96).

If we've used our mouths to teach and bear witness of the truth, to speak kind and uplifting words and to encourage others, then the record kept of our words will justify us and bring great blessings. But if our words have been hurtful or demeaning, they'll bring condemnation upon us. When we come before the Lord in judgment, we'll be stripped of all excuses for the cruel and unworthy things we've said. "Suffer not thy mouth to cause thy flesh to sin; *neither say thou before the angel, that it was an error*" (Eccl. 5:6).

One reason we're held accountable for our spoken words is because they're like a pulse or a temperature reading, indicating the healthiness or illness of our soul. Just as a doctor asks you to open your mouth wide so he can see inside to help with his diagnosis, a person who listens to your words can learn a great deal about your spiritual condition. The words coming *out* of our mouths indicate the condition of what is inside our hearts and minds, "for out of the abundance of the heart the mouth speaketh" (Matt. 12:34). "Out of the same mouth proceedeth blessing and cursing. My brethren, these things ought not so to be" (James 3:10).

Jesus taught that it is "not that which goeth *into* the mouth that defileth a man; but that which cometh *out* of the mouth, *this* defileth a man" (Matt. 15:11). With our modern knowledge of nutrition, we might question the scientific validity of that statement. But Jesus wasn't presenting a technical lesson on nutrition. He was teaching a spiritual principle about the relationship of our character to the words we speak. Peter, who had questions about this, asked Jesus to "declare unto us this parable" (Matt. 15:15). Jesus answered:

> Do not ye yet understand, that whatsoever entereth in at the mouth goeth into the belly, and is cast out into the draught? *But those things which proceed out of the mouth come forth from the heart; and they defile the man.* For out of the heart proceed evil thoughts, murders, adulteries, fornications, thefts, false witness, blasphemies: These are the things which defile a man (Matt. 15:17-20).

As long as we're mortal, fallen, and imperfect, we'll have conflict and war between our righteous thoughts and feelings and the lesser part of us that would speak improper words. But the Lord expects us to control our mouths and our words, just as he expects us to be in control of the rest of our body members. "If any man among you *seem* to be religious, and bridleth not his tongue . . . this man's religion is vain" (James 1:26). Controlling the tongue takes genuine commitment, such as: "I am purposed that my mouth shall not transgress" (Ps. 17:3). "For he that will love life, and see good days, *let him refrain his tongue from evil,* and his lips that they speak no guile" (1 Pet. 3:10). "I will take heed to my ways, that I sin not with my tongue: *I will keep my mouth with a bridle*" (Ps. 39:1).

It may require more time and effort to tame our tongues than it does for any other member of the body. But this we must do if we would be like our Savior. Learning to use our mouths to speak words for righteous influence also takes practice, determination, commitment, and perseverance. Through the power of Jesus Christ and his gospel, we can leave the old man of sin behind and become new creatures, walking *and* talking in the image and will of our Savior. Here's a great resolve: "My lips shall not speak wickedness, nor my tongue utter deceit" (Job 27:4).

The mouth has, of course, other purposes besides speaking. We use our mouths to take in fluids and food to nourish and energize our bodies. While we often take the mouth for granted, it's actually a very complex part of the body. The sense of taste in our tongue, for example, is generated by thousands of complex taste buds. Each taste bud contains twenty-five to forty tiny cells which are only 1/500 of an inch in size. These tiny sensor cells are so fragile and so crucial to our survival that they are replaced with new cells every seven to ten days!

This incredible sense of taste, which is a wonderful gift from our creator, is not for simply adding to the enjoyment of eating. The sense of taste is also vital to our survival, by helping us detect things that are spoiled or not good for us. When we find something in the mouth that tastes rotten, for example, what do we do? Instinctively we spit it out, and this is what we need to do spiritually as well. "But now ye also put off all these; anger, wrath, malice, blasphemy, filthy communications out of your mouth" (Col. 3:8).

USING OUR MOUTHS TO ENCOURAGE OTHERS

"A word spoken in due season, how good is it!" (Prov.15:23).

A man realized he had dialed the wrong phone number when an older woman answered his call. He apologized, but before he could hang up she said, "No, wait! I'm eighty years old, and no one ever calls me. Would you please talk to me for a minute?"

One of the most important things we can do for others is to use our mouth to speak encouragements. "The Lord God hath given me the tongue of the learned, that I should know how to speak a word in season to him that is weary" (Isa. 50:4).

Our words have such powerful impact on others that the mouth probably has more potential to bless (or harm) than any other part of our body. "A soft answer," in response to a verbal attack, for example, "turneth away wrath: but grievous words stir up anger" (Prov. 15:1). Oh "how forcible are right words!" (Job 6:25). Our words can encourage or discourage, lift or pull down, strengthen or weaken the hearers. How blessed one will be when "Thy words have upholden him that was falling" (Job 4:4).

I doubt that anyone who has been uplifted by someone's words—or hurt by them—would doubt that both "Death and life are in the power of the tongue" (Prov. 18:21). The eyes and ears are purely *input* devices for the body, but the mouth is bi-directional; things go *in* and things come out—mostly thoughts and feelings. Of course, the Lord is concerned about the substances which we allow to enter our bodies through our mouths, but the fact that there are more than 14,000 scripture verses on what we should and should not do with our mouths (far more than any other member of the body) certainly testifies to his concern for the expressions that come out of our mouths.

On the positive side, "a wholesome tongue is [nourishing, like] a tree of life" (Prov. 15:4). Those who live righteously, in tune with the scriptures and the Holy Spirit can be a powerful blessing in the lives of others by actually speaking "with the tongues of angels." (See 2 Ne. 31:13-14; 32:2.) In ancient times the survival of a community was contingent on the dependability of their precious well. In like manner, a loving disciple can also preserve life as he speaks words that lift and inspire. "The mouth of a righteous man is a well of life" (Prov. 10:11).

On the other hand, there's probably no other part of the body that can hurt people or get us into trouble as quickly as our mouths. A wicked, cruel tongue is like a weapon of destruction that can strike a victim with an instant wound. "*Their tongue is [like] an arrow shot out*" (Jer. 9:8). Some guns have what is called a "hair trigger," meaning that its actuator has been sensitized for faster action and requires less pressure to fire. Many of our mouths also have "hair triggers," responding and spewing forth words instantaneously, without thought for the effect on the hearers. To be more like the Savior, we must train our mouths to wait for the counsel of our ears before speaking, for "He that answereth a matter before he heareth it, it is folly unto him" (Prov. 18:13). "Seest thou a man that is hasty in his words? there is more hope of a fool than of him" (Prov. 29:20).

Nephi reports that "Laman and Lemuel did speak many *hard words* unto us, their younger brothers" (1 Ne. 3:28). Hard words are cruel. They cut deeply into the emotions. They can quickly destroy feelings of self-worth. They can discourage. They can prompt feelings of spite, hatred, and revenge. They can drive people away from families and the Church. For example, "it came to pass, when I heard these words, that I sat down and wept, and mourned certain days" (Neh. 1:4).

Some people feel so angry with others that they actually ponder and practice hard and cutting words to use against them. They "whet their tongue like a sword, and bend their bows to shoot their arrows, even bitter words" (Ps. 64:3). Some people's tongues are so cruel and skilled in cutting another person down that their tongues are compared to razor blades! "Thy tongue deviseth mischiefs; *like a sharp razor*, working deceitfully" (Ps. 52:2).

Once spoken, our words can't be recalled or erased, any more than you can put toothpaste back in a tube after using it. Thus, scriptures advise us to guard against harming people with our mouths by letting "every man be swift to *hear*, [but] slow to speak, slow to wrath" (James 1:19). "Be not rash with thy mouth, and let not thine heart be hasty to utter anything before God" (Eccl. 5:2).

The Bible records the wicked planning of a verbal attack to discredit a prophet, using nothing but tongues and false words to cut him down and destroy him. "Then said they, Come, and let us devise devices against Jeremiah. . . . . Come, *and let us smite him with the tongue*, and let us not give heed to any of his words" (Jer. 18:18). Have we ever chosen words to cut someone down? Or entered an argument determined to force our own opinion without giving heed to the other's point of view? The Savior's enemies were always trying to trick and trap him into saying something that would justify their opposition. "And as he said these things unto them, the scribes and the Pharisees began to urge him vehemently, and to provoke him to speak of many things: Laying wait for him, and *seeking to catch something out of his mouth, that they might accuse him*" (Luke 11:53-54).

Daniel was unjustly thrown into a den of hungry, ferocious lions because the plotting of his political enemies trapped King Darius into doing it in spite of his favor for Daniel. After fasting the night for Daniel's God to protect him, Darius rushed in the early morning to the den and "cried with a lamentable voice unto Daniel, O Daniel, servant of the living God, is thy God, whom thou servest continually, able to deliver thee from the lions? Then said Daniel unto the king, O king, live forever. My God hath sent his angel, *and hath shut the lions' mouths, that they have not hurt me*" (Dan. 6:20-22). Sometimes the kindest thing we can do when our feelings are hurt is to shut our own mouths so that we don't say things that would hurt others. "Suffer not thy mouth to cause thy flesh to sin" (Eccl. 5:6). "Whoso keepest his mouth and his tongue keepeth his soul from troubles" (Prov. 21:23).

USING OUR MOUTHS TO GOSSIP

"Thou shalt not go up and down as a talebearer among thy people" (Lev. 19:16).

One of the ways Satan led Book of Mormon people into failure and destruction was to "go about spreading rumors and contentions upon all the face of the land, *that he might harden the hearts of the people against that which was good*" (Hel. 16:22). The scriptures contain repeated commandments "to speak evil of no man." (Titus 3:2). Spreading contentions or speaking evil of someone is so destructive to relationships and feelings of self-worth that sowing discord is another of the six things we do with our bodies which the Lord *hates*. (See Prov. 6:19). "Cease to contend one with another; cease to speak evil one of another" (D&C 136:23).

Why does the Lord hate gossip and speaking evil of each other? Because it destroys hope and the desire to move forward. "The words of a talebearer are as wounds, and they go down into the innermost parts of the belly" (Prov. 18:8). This verse describes the wrenching anguish that comes from "a talebearer [who] revealeth secrets" (Prov. 11:13) or who spreads lies and rumors. Sometimes these wounds go so deep they last for years, and may even pass on to succeeding generations. "A brother offended is harder to be won than a strong city" (Prov. 18:19).

How much better our lives could be if we took every opportunity to build and strengthen people with our words rather than gossiping! "Thou shalt not speak evil of thy neighbor, nor do him any harm" (D&C 42:27). Let us rather emulate the Savior by using our mouths to express love, encouragement, and hope.

USING OUR MOUTHS TO SMILE

A utility linesman paralyzed his left side by coming in contact with a live wire. In the lawsuit that followed he was asked to smile before the jury, and because he could only smile on one side of his face, the jury awarded him $100,000 in damages. If half a smile is worth a hundred thousand dollars, what would a whole smile be worth?

One of the kindest things we can do with our mouths is to smile. Can anyone measure or describe the pleasure derived from receiving an unexpected, spontaneous smile from a friend or loved one? Who can describe the pleasure derived by causing an infant to smile? The effect our smiles have on others is so important that newscasters and other public figures are actually trained to smile in a pleasing way.

A smile can disarm another person's anger and aggression. I knew a man named Ted whose management position required him to participate in weekly accountability meetings at our aerospace manufacturing plant. These meetings were known for the bloodshed they caused. Instead of the department heads working together to meet production schedules, the meetings were characterized by self-defensive blame and backstabbing. Shouting at each other was a frequent occurrence. Then Ted, who was not even in charge of the meetings, changed everything without saying a word. He attended a personal development course where he learned to smile. He took this new skill back to the hostile meetings, and when people attacked his department he no longer retaliated with words but just sat and smiled at them. Before long their anger melted. After all, who can shout at a smiling face? When his opponents calmed down, he was able to discuss the issues and resolve the problems. It was infectious.

I also remember a sales trainer who had a winning smile. He would take us cold-calling from business to business and charge in to an office almost shouting, "Who's in charge here? I want to sell him something!" This invariably aroused hostility and verbal abuse from the busy manager, which just tickled Marlow. He stood there grinning from ear to ear, and when the offended manager realized his anger was just melting on this huge smile, he couldn't help calming down and listen to what was being offered. The power of this man's smile was so amazing that he got an audience in almost every office.

The Lord has placed thirty muscles in our faces to produce smiles and the accompanying facial expressions. There are separate muscles for the mouth, eyebrows, and eyes. Knowing the pleasure we receive from freely given smiles, can we not be more Christlike in the attempt to bestow frequent smiles on others?

Two years before Melvin J. Ballard was called as an apostle, he met the Savior face to face in a dream in which he was taken into the Salt Lake Temple. The thing about this visit that impressed him most was the divine smile he received from the Savior. He said:

I was led into a room where I was informed I was to meet someone. As I entered the room I saw, seated on a raised platform, the most glorious being I have ever conceived of, and was taken forward to be introduced to Him. As I approached He smiled, called my name, and stretched out His hands toward me. *If I live to be a million years old I shall never forget that smile (Melvin J. Ballard: Crusader for Righteousness* [Salt Lake City: Bookcraft Inc., 1966], p. 66).

When I'm tempted to sin, remembering that approving smile helps me to make the right choices. As we work to improve our spiritual anatomy, we would be wise to frequently ponder our relationship with the Savior and whether our prayers and lives would prompt him to smile upon us with approval.

USING OUR MOUTHS TO FURTHER HIS WORK

The Lord used his mouth every day of his ministry to declare the truth and teach people how to have happier lives. Our duty as disciples of Christ is to reflect him not only in our deeds, but also in our words as we "let [our] conversation be as it becometh the gospel of Christ" (Philip. 1:27). He expects us to speak for him. "It is necessary and expedient in me that you should open your mouths in proclaiming my gospel, the things of the kingdom" (D&C 71:1; also 30:5).

"Let no corrupt communication proceed out of your mouth," Paul advised, but learn to use it for "that which is good to the use of edifying, that it may minster grace unto the hearers" (Eph. 4:29). Teaching and witnessing for Christ is not to be an occasional duty for his disciples, but a continual readiness and lifetime commitment. "But sanctify the Lord God in your hearts and *be ready always to give an answer to every man* that asketh you a reason of the hope that is in you with meekness and fear" (1 Pet. 3:15). "And *thou must open thy mouth at all times,* declaring my gospel with the sound of rejoicing" (D&C 28:16).

Many of us fear the obligation of teaching or speaking in church. This natural fear might be legitimate if the success of our service depended on the skills of our own mouths. But when we speak in God's service, we're entitled to have our words quickened by his Spirit, so that we say things that touch hearts and change lives in ways we might not even have thought of ourselves. "Open thy mouth, and it shall be filled, *and I will give thee utterance*" (Moses 6:32). How does he give us "utterance"? Through the Holy Ghost. "And it shall come to pass, that if you shall ask the Father in my name, in faith believing, you *shall* receive the Holy Ghost, *which giveth utterance*, that you may stand as a witness . . . and also that you may declare repentance unto this generation" (D&C 14:8).

The Lord understands that we're sometimes timid and lacking in confidence. We therefore hesitate to speak, teach, or testify on his behalf. But he said, "You shall ever open your mouth in my cause, not fearing what man can do, *for I am with you*" (D&C 30:11). Knowing that he is with us can give us the confidence to receive his grace and enabling power for our words.

It's displeasing to the Lord when disciples who have made sacred covenants with him fail to use their mouths in his service. "But with some I am not well pleased, for they will not open their mouths, but they hide the talent which I have given unto them, because of the fear of man. Wo unto such, for mine anger is kindled against them" (D&C 60:2). On the other hand, he has great blessings ready for those who do use their mouths in his service. "Yea, open your mouths and spare not, and you shall be laden with sheaves upon your backs, for lo, I am with you" (D&C 33:9).

Many reluctant prophets have recorded their hesitancy to speak for the Lord and the endowment of grace they received when they submitted to his will. Jeremiah, for example, resisted his prophetic calling by saying, "Ah, Lord God! behold, I cannot speak: for I am a child" (Jer. 1:6). But the Lord assured him that if he was willing to serve, he would be with Jeremiah and give him the words to say. Jeremiah then reports that "the Lord put forth his hand, and touched my mouth. And the Lord said unto me, Behold, I have put my words in thy mouth" (Jer. 1:9). Few prophets have been as powerful in their testimonies as Jeremiah.

Moses became an outstanding landmark in the panorama of valiant prophets, but he started as most of God's servants do: timid and acutely aware of his own limitations. "But, behold, they will not believe me, nor hearken unto my voice: for they will say, The Lord hath not appeared unto thee" (Ex. 4:1). The Lord reasoned with him for the next eight verses and taught him several miracles he could perform to impress the superstitious people. Moses was still afraid and said, "O my Lord, I am not eloquent, neither heretofore, nor since thou hast spoken unto thy servant: but I am slow of speech, and of a slow tongue" (Ex. 4:10).

Doesn't this sound like the excuses *we* make when trying to dodge a teaching or speaking assignment? "And the Lord said unto him, "who hath made man's mouth? . . . have not I the Lord? Now therefore go, and *I will be with thy mouth*, and teach thee what thou shalt say" (Ex. 4:11-12). The Lord doesn't need our *capability* to fulfill his work as much as our *availability*. Throughout history he has manifested his power to make us equal to our assignments once we agree to serve willingly and place ourselves in his hands.

HONEST LIPS OR HYPOCRITICAL LIPS

"The wicked is snared by the transgression of his lips" (Prov. 12:13).

In the scriptures, lips symbolize hypocrisy. Speaking of the false churches and ministers, the Lord told Joseph Smith, "*They draw near to me with their lips*, but their hearts are far from me" (JS-H 1:19; see also Isa. 29:13; Matt. 15:8; 2 Ne. 27:25). "They bless with their mouth, but they curse inwardly" (Ps. 62:4). "Behold, O God, they cry unto thee, and yet their hearts are swallowed up in their pride. Behold, O God, they cry unto thee with their mouths, while they are puffed up, even to greatness, with the vain things of the world" (Alma 31:27).

One of the tenderest, most intimate things we can do with our lips is to kiss. A kiss can express casual friendship, family love, passionate feelings, or respect, as with those who have had the privilege of kissing Christ's feet. (See Luke 7:38, Ne. 11:19; 17:10.)

The most famous kiss in history was that which Judas used to betray the Savior. "Now he that betrayed him gave them a sign, saying, Whomsoever I shall kiss, that same is he: hold him fast" (Matt. 26:48). Judas led the mob to the Garden where Christ had been working out the atonement the night before his crucifixion and "he came to Jesus, and said, Hail, master; and kissed him" (Matt. 26:49). Truly Judas "drew near" to the Lord with his lips, but his heart and loyalty were far from him. "Jesus said unto him, Judas, betrayest thou the Son of man with a kiss?" (Luke 22:48). It's easy to condemn Judas, as we hastily proclaim that *we* would never do what he did. But betraying lips are an epidemic in our society, as husbands and wives betray their sacred marriage vows by sharing kisses and other intimacies with companions other than the one to whom they were promised. When we kneel to pray, are we doing so with clean lips, with honorable lips, or with lips that have betrayed another?

". . . but he that refraineth his lips is wise" (Prov. 10:19).

USING OUR MOUTHS TO LIE

"And again, the Lord God hath commanded that men. . . should not lie" (2 Ne. 26:32).

The consequences of deliberate deceit and falsehood on both the perpetrator and the recipient can be so devastating that "a lying tongue" and "a false witness that speaketh lies" are listed among the six things which the Lord hates. (See Prov. 6:16-19.) "Lying lips are abomination to the Lord; but they that deal truly are his delight" (Prov. 12:22).

The commandment to tell the truth and prevent our mouths from lying has been expressed throughout the history of God's instructions to man. From "Thou shalt not bear false witness" in the original ten commandments (Ex. 20:16) to the directive in our time of restoration "Thou shalt not lie; he that lieth and will not repent shall be cast out" (D&C 42:21), the message is the same.

"Lie not one to another," said Paul, "seeing that ye have put off the old man with his deeds" (Col. 3:9). "Put away lying," he emphasized, and "speak every man truth with his neighbor" (Eph. 4:25; also Zech. 8:16). The Bible speaks of those who have practiced at lying so much that "they have *taught* their tongue to speak lies" (Jer. 9:5) "and their tongue is deceitful in their mouth" (Micah 6:12). Can you imagine trying to get your prayers through when the "caller ID" up there flashes, "Incoming prayer from a lying tongue"? "Is there iniquity in my tongue?" (Job 6:30).

Jesus was the messenger of truth. When we speak the truth, we emulate and align ourselves with him. When we use our mouths to lie and deceive, we emulate the one who represents everything that is evil and align ourselves with the devil, described as "the father of all lies" (Ether 8:25; 2 Ne. 2:18; 9:9). "When he speaketh a lie, he speaketh of his own: for he is a liar, and the father of it" (John 8:44). God, on the other hand, the very symbol of perfection and truth, *"will not* lie" (1

Sam. 15:29); he *"cannot* lie" (Titus 1:2; D&C 62:6; Enos 1:6; Ether 3:12); indeed, it is *"impossible* for God to lie" (Heb. 6:18). Likewise, "the Spirit speaketh the truth and lieth not" (Jacob 4:13).

It isn't likely that we'd lie to someone unless we thought we'd be successful in deceiving them. A skillful liar may deceive those who know him well—for a while. But there's one person we can never deceive, and that is God, "for he cannot be deceived" (2 Ne. 9:41). *"Do ye imagine to yourselves that ye can lie unto the Lord* in that day, and say—Lord, our works have been righteous works upon the face of the earth—and that he will save you?" (Alma 5:17).

Truth is open and revealing with no secrets. Lying can be as simple as concealing something people have a right to know. Sometimes silence, a deliberate failure to correct a false impression, is as much a lie as saying words that are not true. But this, too, is sin, for "it is contrary to the order of heaven for a just man to deceive" (D&C 129:7).

Because the Lord has granted us agency, a mouth that lies is allowed to continue doing so as long as the person wishes, but every lie carries a penalty. Every word spoken is recorded, and the day of accountability will surely come, for "a false witness shall not be unpunished, and he that speaketh lies shall not escape" (Prov. 19:5). "Verily, verily, I say unto you, wo be unto him that lieth to deceive . . . for such are not exempt from the justice of God" (D&C 10:28).

"But wo unto them that are deceivers and hypocrites, for, thus saith the Lord, I will bring them to judgment" (D&C 50:6). The consequences of continued and deliberate lying are well defined and certainly worth avoiding. "Wo unto the liar, for he shall be thrust down to hell" (2 Ne. 9:34). "Wherefore, I, the Lord, have said that . . all liars, and whosoever loveth and maketh a lie . . . shall have their part in that lake which burneth with fire and brimstone, which is the second death" (D&C 63:17).

Jesus said he would take personally our efforts to serve (or lack of effort) as a measure of our judgment. "Inasmuch as ye *have done it* unto one of the least of these my brethren, ye have done it unto me," he said, and "Inasmuch as ye *did it not* to one of the least of these, ye did it not to me" (Matt. 25:40, 45). This principle also applies to the words we speak. When we lie to each other, we're also lying to God. We see this principle in the words of Alma, who caught a famous lawyer in his deliberate lies: "Now Zeezrom, seeing that thou hast been taken in thy lying and craftiness, *for thou hast not lied unto men only but thou hast lied unto God*; for behold, he knows all thy thoughts, and thou seest that thy thoughts are made known unto us by his Spirit" (Alma 12:3; see also Acts 5:4).

The consequences of lying are not all postponed until the day of judgment. I know a man who sent a package to a foreign country, which required a custom form stating the contents and the value. To reduce the cost of import duties, he lied about the value. The ship sank. When his damaged package was retrieved, he only received reimbursement for the low value he had stated instead of its true value.

If you find yourself "possessed with a lying spirit" (Alma 30:42), you might consider this honest prayer for help in controlling your mouth: "Deliver my soul, O Lord, from lying lips, and from a deceitful tongue" (Ps. 120:2).

USING OUR MOUTHS TO BOAST

"If I justify myself, mine own mouth shall condemn me" (Job 9:20).

The specific commandments against boasting are typified by these two verses: "See that ye are not lifted up unto pride; yea, see that ye do not boast in your own wisdom, nor of your much strength" (Alma 38:11), "neither boast of faith nor of mighty works" (D&C 105:24).

Using our mouths to boast is a symptom of isolation from the Lord. It indicates self-centeredness and pride instead of recognition and appreciation for his influence in our lives. As Samuel the Lamanite prophet said, "Ye do not remember the Lord your God in the things with which he hath blessed you, but ye do always remember your riches [or personal achievements], not to thank the Lord your God for them; yea, *your hearts are not drawn out unto the Lord*, but they do swell with great pride, unto boasting, and unto great swelling" (Hel. 13:22). The Lord has warned that such boasting will isolate us from his blessings. We'll be left to ourselves to learn just how insufficient we are without his divine influence and strength in our lives. "And because of . . . their boastings in their own strength, *they were left in their own strength*; therefore they did not prosper, but were afflicted and smitten, and driven before the Lamanites, until they had lost possession of almost all their lands" (Hel. 4:13). "For although a man may have many revelations, and have power to do many mighty works, *yet if he boasts in his own strength*, and sets at naught the counsels of God, and follows after the dictates of his own will and carnal desires, he must fall and incur the vengeance of a just God upon him (D&C 3:4).

King Benjamin gave a great discourse in which he included the silly pointlessness of boasting. He told his people that all the Lord asked of them was to trust him and obey him, "for which if ye do, he doth immediately bless you; and therefore he hath paid you. And ye are still indebted unto him, and are, and will be, forever and ever; *therefore, of what have ye to boast?*" (Mosiah 2:24).

Boasting words, like all other words we speak, are always heard and recorded above. "Thus with your mouth ye have boasted against me, and . . . *I have heard them*" (Ezek. 35:13). "Talk no more so exceeding proudly," is the counsel of another Samuel. "Let not arrogancy come out of your mouth: for the Lord is a God of knowledge, and by him actions are weighed" (1 Sam. 2:3).

There's one thing, however, in which we're encouraged to boast, and that is in the majesty and goodness of our God. "In God we boast all the day long, and praise thy name for ever" (Ps. 44:8). Ammon was praising God so exuberantly that his brother, Aaron, rebuked him, saying: "Ammon, I fear that thy joy doth carry thee away unto boasting" (Alma 26:10). Ammon's response teaches us a lot about the appropriateness of using our mouths to praise God.

> And Ammon said unto him: I do not boast in my own strength, nor in my own wisdom; but behold, my joy is full, yea, my heart is brim with joy, and I will rejoice in my God. Yea, I know that I am nothing; as to my strength I am weak; *therefore I will not boast of myself, but I will boast of my God,* for in his strength I can do all things (Alma 26:11-12).

USING OUR MOUTHS TO EXPRESS PRAISE

"Oh that men would praise the Lord for his goodness, and for his wonderful works to the children of men!" (Ps. 107:8).

Praising God and his goodness is a specific commandment that has been repeated many times. "Praise the Lord, call upon his name, declare his doings among the people, make mention that his name is exalted" (Isa. 12:4). "Ye are a chosen generation," said Peter, "a royal priesthood, an holy nation, a peculiar people; that *ye should shew forth the praises* of him who hath called you out of darkness into his marvelous light" (1 Pet. 2:9). One definition of praise is to simply "make known his deeds among the people" and to "talk ye of all his wondrous works" (1 Chr. 16:8-9). "And we talk of Christ, we rejoice in Christ, we preach of Christ, we prophesy of Christ, and we write according to our prophecies, that our children may know to what source they may look for a remission of their sins" (2 Ne. 25:26).

Praising God is an indicator of the reverence, awe, respect, and appreciation we feel for our Savior and Heavenly Father. There will be no praise if we're empty and devoid of these spiritual feelings. Each Sunday, during the sacrament service, we pledge to do our best to *remember* the Savior and our covenants with him during the competing affairs of our daily life. Our mouths can help our memories by developing the habit and attitude of continual praise. "Let my mouth be filled with thy praise and with thy honour all the day" (Ps. 71:8). "From the rising of the sun unto the going down of the same the Lord's name is to be praised" (Ps. 113:3).

Our morning and evening prayers should include specific words of praise, as we kneel "every morning to thank and praise the Lord, and likewise at even" (1 Chr. 23:30). "My tongue also shall talk of thy righteousness all the day long" (Ps. 71:24). If we would use our mouths as instruments of righteousness to direct our attention away from the rotten things we face in the world each day, we could make this pledge every morning: "I will bless the Lord at all times: his praise shall continually be in my mouth" (Ps. 34:1). "And my tongue shall speak of thy righteousness and of thy praise all the day long" (Ps. 35:28).

There are specific *places* in which we've been asked to utter praises to our God. One of them is in the buildings that have been dedicated to his worship. "Praise God in his sanctuary" (Ps. 150:1). Attending church or the temple should involve more than just being there. Our hearts should be meditative and pondering his goodness to us and uttering praises to him. "Enter into his gates with thanksgiving, and into his courts with praise: be thankful unto him, and bless his name" (Ps. 100:4). "I will declare thy name unto my brethren: in the midst of the congregation will I praise him" (Ps. 22:22).

The scriptures also list specific *subjects* that we're commanded to include in our praise. "Who can utter the mighty acts of the Lord? who can shew forth all his praise?" (Ps. 106:2). When we contemplate the wonders of creation, from the spectacular endless reaches of the galactic heavens to the finite structure of a molecule, surely our mouths should praise the wonder of his greatness. "For great is the Lord, and greatly to be praised" (1 Chr. 16:25). "Praise him for his mighty acts: praise him according to his excellent greatness" (Ps. 150:2).

Have you ever contemplated the perfect goodness of our Heavenly Father and Savior? Have you ever thought about just how hard it is to anger them or turn them away from our best interests? Or how eager they are to welcome us back when we repent? Is this not a major reason to use our mouth in praise? "I will praise the Lord according to his righteousness," for "he is good; for his mercy endureth for ever" (Ps. 7:17; 2 Chr. 7:3). When we really love and appreciate the Lord, we should find ourselves compelled to worship and praise him simply because he "is worthy to be praised " (2 Sam. 22:4; Ps. 18:3). "Seven times a day do I praise thee because of thy righteous judgments" (Ps. 119:164).

One of the easiest things to praise God for is the blessings he gives us. "I will mention the lovingkindnesses of the Lord, and the praises of the Lord, *according to all that the Lord hath bestowed on us*" (Isa. 63:7). "O Lord, thou art my God: I will exalt thee, I will praise thy name; for thou hast done wonderful things" (Isa. 25:1).

But real worship, real devotion, is expressed when we can praise God even during our pain, even when terrible things are happening to us, when things look and feel dark and gloomy. There's something majestic and divine in the person whose mouth can praise God in the very midst of affliction, pain, and sorrow. My wife read of a prisoner of the Nazis who taught her sister to express thanks and praise God for every torture and deprivation they suffered, even for the fleas that tormented them in their barracks. It was hard to do until they discovered that the women in their barracks, because of those fleas, were the only ones in the camp not used for the sexual pleasures of the guards. (Corrie Ten Boom, *The Hiding Place* [Uhrichsville, Ohio: Barbour Publishing, Inc., n.d.], pp. 195-204.)

We stand in awe at the faithfulness of Nephi. His family murmured because of their sufferings and turned on him in bitter persecution and ridicule, even seeking his life. "Nevertheless," he said, "I did look unto my God, and I did praise him all the day long, and I did not murmur against the Lord because of mine afflictions" (1 Ne. 18:16).

The Anti-Nephi-Lehi converts were attacked and slaughtered by their former Lamanite friends. And yet they were so grateful to God for rescuing them from the false traditions of the Lamanites that they "would lie down and perish, and praised God even in the very act of perishing under the sword" (Alma 24:23). "Therefore let us offer the *sacrifice of praise* to God continually, that is, the fruit of our lips giving thanks to his name" (Heb. 13:15). It's when we can love and praise God in spite of those circumstances that we offer the "sacrifice of praise" and bring miracles into our lives. (See Heb. 13:15.) "For this is thankworthy, if a man for conscience toward God endure grief, suffering wrongfully. For what glory is it, if, when ye be buffeted for your faults, ye shall take it patiently? but if, when ye do well, and suffer for it, ye take it patiently, this is acceptable with God" (1 Pet. 2:19-20).

USING OUR MOUTHS TO EXPRESS GRATITUDE

"And be ye thankful" (Col. 3:15).

It's been said that "thank you" are two of the most powerful words anyone can use. Expressing our thanks to God (and to others) is one of the most important things we can do with our mouths. That's why there are more than two hundred scriptures on this subject. President Marion G. Romney said, "To thank the Lord in all things is not merely a courtesy, it is a commandment as binding upon us as any other commandment" (*Ensign*, Nov. 1982, p. 50). One characteristic of Moroni that the Book of Mormon praises is the fact that he was "a man *whose heart did swell with thanksgiving* to his God, for the many privileges and blessings which he bestowed" (Alma 48:12).

"*O how you ought to thank your heavenly King*," said King Benjamin, who then went on to explain that because we've been given so much, it would be impossible for us to ever express sufficient gratitude, even "if you should render all the thanks and praise which your whole soul has power to possess" (Mosiah 2:19-20).

Apostle Bruce R. McConkie stated: "True worship includes thanksgiving to God—the acknowledging and confessing with joy and gladness of the benefits and mercies which he bestows upon his children" and that "among all sins, none is so prevalent as the sin of ingratitude" (*Mormon Doctrine* [Salt Lake City: Bookcraft, 1966], pp. 788, 380). Elder Neal A. Maxwell said, "Ingratitude reflects the intellectual dishonesty of those who can enumerate their grievances but cannot count their blessings" (*Notwithstanding My Weaknesses* [Salt Lake City: Deseret Books, 1981], p. 53). We find very few things in the scriptures that we can do to actually offend God, but failing to notice and appreciate his influence in our lives is one of them. (See D&C 59:20-21.)

Isaiah spoke of those who are in danger of offending God because they're so preoccupied with the things of this world that "they regard not the work of the Lord, *neither consider* [notice and appreciate] *the operation of his hands*" (Isaiah 5:12). It's interesting that the very next verse says, "*Therefore* my people are gone into captivity, because they have no knowledge" (Isa. 5:13). This statement usually serves to substantiate the need for truth in the process of salvation. But when we link it to the preceding verse, we learn that preoccupation with the affairs of mortal life, which prevent us from noticing and appreciating God's hand in our daily lives, robs us of the knowledge we need to draw close to God and to be like him.

"It is a good thing to give thanks unto the Lord" (Ps. 92:1). Why is it so good? Not because *God* needs our thanks, but because *we* need to be closer to him. The sincere expression of gratitude makes that possible, while failing to notice and thank God for his daily blessings isolates us from him and pushes us more into the affairs of the world.

Gratitude opens the door to divine fellowship. We can't be close to the Lord until we learn to appreciate his divine character and what he's trying to do in our lives. There's an actual spiritual formula here: the *more* we feel and express our gratitude, the closer God will feel; the *less* grateful we are, the farther away he will seem. Thus Amulek taught that we should "live in thanksgiving *daily*, for the many mercies and blessings which he doth bestow upon [us]" (Alma 34:38). Daniel had such a powerful habit of gratitude that "he kneeled upon his knees three times a day, and prayed, and gave thanks before his God" (Dan. 6:10).

In addition to "living in thanksgiving *daily*," there are two specific times during the day when we're instructed to emphasize the expression of our thankfulness. The first is in the morning, when we should awake with feelings of gratitude to our God for another day in which we can practice becoming more like him: "And when thou risest in the morning let thy heart be full of thanks unto God" (Alma 37:37). Second, when we retire for the evening we're specifically instructed to ponder the events and blessings of the day. This is a great time to look back and discern the places and events of the day that might reveal God's hands in our affairs, so that we not only "pray unto him continually by day," but also "*give thanks unto his holy name by night*" (2 Ne. 9:52). As we kneel to say our evening prayers, "Let us come before his presence with thanksgiving" (Ps. 95:2).

In addition to teaching our mouths to give thanks to God at *all times*, we're instructed to be so watchful for his influence in our lives that we can notice and express gratitude for *every* blessing we receive, "always returning thanks unto God *for whatsoever things ye do receive*" (Alma 7:23). "And ye must give thanks unto God in the Spirit *for whatsoever blessing* ye are blessed with" (D&C 46:32).

The path that takes us closer to God will always be a path cluttered with tutoring trials and tribulations. Even Christ had to suffer difficult lessons, but did so willingly because that is the way we learn obedience. (See Heb. 5:8.) The knowledge of this essential process should help us to be more willing and grateful to walk in his steps, no matter how difficult the journey. It's easy to thank God for *blessings*; even the most casual disciple can do that. The biggest challenge in sending our gratitude to God is not in the *frequency* of our gratitude, but in learning to thank him for *everything* that happens, just as we're to praise him for everything—even when our limited vision interprets it as damaging and harmful. This is the area where the prophets have placed the most emphasis, for this emotional Mount Everest is where we'll obtain the most growth.

Jesus talked about how easy it is to respond positively when good things happen, but we need to grow to the point where we can always respond with a positive attitude, *independent* of the perceived merit of the circumstances. "For if ye love them which love you, what thank have ye?" he asked, "for sinners also love those that love them" (Luke

6:32). Or, "if ye do good to them which do good to you, what thank have ye? for sinners also do even the same" (Luke 6:33).

We go to the doctor for a shot, knowing it's going to hurt for a moment, but grateful for the immunity it will give us. It's the same with unwanted afflictions. Lehi said, "*he shall consecrate thine afflictions for thy gain*" (2 Ne. 2:2). And Jesus said of the afflictions we must learn to appreciate, "all things wherewith you have been afflicted shall work together for your good" (D&C 98:3). Having such knowledge and confidence in the value of the disturbing lessons we would otherwise resent will allow us to say with Paul, "We are troubled on every side, yet not distressed; we are perplexed, but not in despair" (2 Cor. 4:8).

Jesus also said that no matter what bad things happen to you, "Let not your heart be troubled: ye believe in God, believe also in me" (John 14:1). What power can keep us from feeling "troubled" when bad things happen to us? Our trust in God. The principle we must keep in mind is that God has promised to "order all things for [our] good, *as fast as* [we] *are able to receive them*" (D&C 111:121). If we can't feel grateful for the circumstances, we can at least be grateful that he's in charge of those circumstances and is guiding them for our best good.

We'd certainly rejoice and give thanks at the discovery that some benefactor had deposited a great sum of money in our bank account. Every adversity, every injustice and affliction is like a deposit in our account, which will most certainly result in compensatory blessings— if not now, then later. Paul expressed what we already know about these deposits, that "no chastening for the present seemeth to be joyous, but grievous." Nevertheless, he said, and this is the part to which we cling for our present expressions of gratitude, "*afterward* it yieldeth the peaceable fruit of righteousness" but only to the people, he said, "which are *exercised* thereby" (Heb. 12:11). The Greek translation for the word "exercised," as used in this verse, is to be "trained and disciplined."

Paul promised that *all things* can work together for good to them that love God (see Rom. 8:28), but we must realize that all things do not necessarily "work together for good" to those who refuse to be "exercised," or who respond negatively and impatiently. When we get all tangled up in the seeming injustices of our trials, trying to blame God (or someone else) for the very situations that allow us to learn and grow, we negate the benefits and experiences they were intended to provide. It's like rejecting that bank deposit made to our spiritual account.

As we make our way through mortality, being "trained, disciplined, and exercised" by the lessons of life, we encounter many oppositions and disappointments that test our trust in God and his divine purposes. We may respond to these difficult circumstances with gratitude for the learning opportunity they provide, or with negative emotions such as doubt and resentment—the "Why-is-God-doing-this-to-me?" attitude. Knowing that we shall "reap eternal joy for all our sufferings" (D&C 109:76) should definitely improve our attitude.

When we were in school, we grew weary of the endless homework assignments and wished there weren't so much studying and work to do. But we realized it was worth it when we received that diploma. Now we don't even remember the frustrations of the individual classes and assignments. So it will be with our mortal schooling. We struggle with the day-to-day disappointments, heartaches, and adversities, but they all add to our knowledge and experience and contribute to our becoming more Christlike and qualified to graduate to eternal life. And when we stand before him and Heavenly Father and receive that "Well done, thou good and faithful servant" diploma, we'll no longer care about the individual pains that helped us learn and grow and become what we needed to be.

Our challenge, then, in facing difficulties is to learn to become "as a child, submissive, meek, humble, patient, full of love, *willing to submit to all things* which the Lord seeth fit to inflict upon him, even as a child doth submit to his father" (Mosiah 3:19). Resenting the opposition and tutoring circumstances we came here to experience is like paying a huge tuition to go to college and then ditching or resenting the classes. We can't be willing or grateful while we're murmuring and complaining over our discomforts and problems.

Paul also used the compelling words "all things" when he said we should be "giving thanks always *for all things* unto God and the Father in the name of our Lord Jesus Christ" (Eph. 5:20; also D&C 59:7). He also said, "In *every thing* give thanks: for this is the will of God in Christ Jesus concerning you" (1 Thess. 5:16). We notice Paul did not say "everything," but "every thing" and "all things." What's the difference? The word "everything" means a collection of everything in a group. For example, we may be thankful for the *principle* of opposition that helps us choose and grow, but have difficulty being thankful for the individual parts of that opposition. Opposition and adversity as an abstract category, for example, doesn't feel the same as the death of a loved one, the loss of a job, or the smearing of our reputation—specific, individual difficulties.

Since the Bible uses the word "everything" in other verses, I believe Paul's careful use of the words "every thing" and "all things" was his way of teaching us to express gratitude for each painful ingredient in the whole curriculum of mortal schooling. Christ has said that the person who learns to receive "*all things* with thankfulness shall be made glorious" (D&C 78:19). Brigham Young also said, "*Every trial and experience you have passed through is necessary for your salvation*" (as quoted in *My Errand From the Lord*, 1976-77 Melchizedek Priesthood manual [Salt Lake City: The Church of Jesus Christ of Latter-day Saints, 1976], p. 228). Surely these are good reasons to rejoice and give thanks for every event in our lives, even when they don't seem good at the moment.

It's certainly not easy to express gratitude for adversity. Many times I have to pray, "Heavenly Father, I don't like this situation. It hurts and I don't understand it, but I thank thee for it. I'm asking thee to help me find the good in it, as the scriptures promise. Please show me why I need this lesson. What are you trying to teach me?" He doesn't always answer that prayer as quickly or directly as I'd like. I think sometimes he withholds the answer because it's more important for me to learn trust and patience than it is to immediately overcome the difficult circumstance. And if that's what I need to learn and develop, shouldn't I be thankful he cares enough about me to make sure that I get the necessary homework?

For this is thankworthy, if a man for conscience toward God endure grief, suffering wrongfully. For what glory is it, if, when ye be buffeted for your faults, ye shall take it patiently? but if, when ye do well, and suffer for it, ye take it patiently, this is acceptable with God. For even hereunto were ye called: because Christ also suffered for us, leaving us an example, that ye should follow his steps (1 Pet. 2:19-21).

We can learn to sincerely express thankfulness for our troubling circumstances when we know that adversity doesn't happen *to* us—it happens *for* us.

USING OUR MOUTHS TO MURMUR

"Do all things without murmurings and disputings" (Philip. 2:14).

We use our mouths to put nourishment into our bodies. The food we eat produces waste materials, which come out of other parts of the body. Unfortunately, a lot of trash comes out of our mouths as well. Perhaps the opposite of expressing feelings of gratitude and praise with our mouths is the ingratitude and lack of trust we manifest when we murmur.

To "murmur" is to complain and grumble: to express feelings of dissatisfaction, resentment, and discontent. How different this behavior is from that of Job, who certainly deserved to complain over his injustices, but said, instead, "Though he slay me, yet will I trust in him" (Job 13:15). Murmuring is a way of saying that our opinions have greater value than the opinions of those we're criticizing. It's very dangerous to our spiritual welfare, because the Savior warned, "Whosoever shall say, Thou fool, shall be in danger of hell fire" (3 Ne. 12:22). "And when the people complained, it displeased the Lord: and the Lord heard it; and his anger was kindled" (Num. 11:1).

Every time something went wrong for the people of Moses, they complained against him, as if bondage in Egypt would have been preferable to the difficulties of their wanderings in the wilderness. "*The Lord heareth your murmurings which ye murmur against him,*" Moses warned, but "your murmurings are not against us, but against the Lord" (Ex. 16:8). "And the Lord spake unto Moses and unto Aaron, saying, How long shall I bear with this evil congregation, which murmur against me? *I have heard the murmurings of the children of Israel,* which they murmur against me" (Num. 14:26-27).

In the Bible, the ungrateful Israelites were often condemned because they "murmured in [their] tents" (Deut. 1:27; Ps. 106:25). Many times when *we* murmur and complain, we also do it at home. Murmuring at home will cause our children, having no opposite view to judge by, to accept our ideas as truth and also to feel that it must be okay to demean others. We may think our murmuring will go unnoticed if it's done in the privacy of "our tents," but the Lord is aware of every word spoken by our mouths. "And Moses spake unto Aaron, Say unto all the congregation of the children of Israel, Come near before the Lord: *for he hath heard your murmurings*" (Ex. 16:9). "Therefore whatsoever ye have spoken in darkness shall be heard in the light; and that which ye have spoken in the ear in closets shall be proclaimed upon the housetops" (Luke 12:3).

A murmuring attitude of discontent and lack of trust is certainly not conducive to the spiritual growth or encouragement of those around us. "For verily, verily I say unto you, he that hath the spirit of contention is not of me, but is of the devil, who is the father of contention" (3 Ne. 11:29). Murmuring is displeasing to the Lord, not only for its destructive influence on *others*, but also for the damage it does to *our own* personal relationship with him. "I, the Lord, chasten him for the murmurings of his heart" (D&C 75:7).

We're familiar with the unceasing murmurings of Laman and Lemuel. We marvel how they could murmur immediately after being visited by an angel. "And after the angel had departed, *Laman and Lemuel again began to murmur*, saying: How is it possible that the Lord will deliver Laban into our hands?" (1 Ne. 3:31). Laman and Lemuel murmured because they couldn't see a divine purpose in their difficulties. They didn't know or trust the good will of their God. "And they did murmur because they knew not the dealings of that God who had created them" (1 Ne. 2:12).

Think back to the last few things that caused you to murmur or complain. See if you can discover a relationship with Laman and Lemuel's lack of vision. It's a profound truth that *what holds our attention holds us*. It's difficult to complain about our difficulties when we first remember and appreciate all the counterbalancing blessings in our lives. Nephi explained that this is exactly how he kept his mind off his difficulties and maintained his unwavering faith. "Nevertheless, *I did look unto my God, and I did praise him all the day long*; and [therefore] I did not murmur against the Lord because of mine afflictions" (1 Ne. 18:16). "Therefore go, my son, and thou shalt be favored of the Lord, because thou hast not murmured" (1 Ne. 3:6).

If anyone ever had the right to complain or murmur because of unjust treatment, it would have been the Savior. But in this, as in all things, he set the example for us in the way we should control our mouths. "He was oppressed, and he was [unjustly] afflicted, *yet he opened not his mouth*: he is brought as a lamb to the slaughter, and as a sheep before her shearers is dumb, *so he openeth not his mouth*" (Isa. 53:7). "Although in agony he hung, No murmuring word escaped his tongue" (*Hymns*, 191).

GNASHING OUR TEETH

I once met a man with a mouth full of chrome teeth! After all his defective teeth had been pulled, he felt the dentists wanted too much money to make him a set of false teeth, so this farmer made his own— out of the chrome bumper of an abandoned car!

The hardest substance in the body is tooth enamel. In fact, teeth are harder than bones. But surprisingly, the teeth are considered a specialized part of the skin rather than bone! Perhaps that adds meaning to the phrase, "I am escaped with the skin of my teeth" (Job 19:20). Tooth enamel doesn't replenish or replace itself, so a person with 100-year-old teeth still has the same enamel he got in the womb. Perhaps this long-lasting feature of the teeth is why the predominant use of teeth in scripture is to denote the anguish and remorse that will be suffered by those who delay their repentance until it's everlastingly too late.

Who could describe the horrible feelings you would have to see your friends and family members, your own parents, brothers and sisters, your children and spouse welcomed into the celestial glory for eternity without you? But because of the way you chose to live your life, you were not qualified to go with them and would be forever confined to another, lesser realm. "There shall be weeping and gnashing of teeth, when ye shall see Abraham, and Isaac, and Jacob, and all the prophets, in the kingdom of God, *and you yourselves thrust out*" (Luke 13:28). "And then shall the wicked be cast out, and they shall have cause to howl, and to weep, and wail, and gnash their teeth; *and this because they would not hearken unto the voice of the Lord*; therefore the Lord redeemeth them not" (Mosiah 16:2).

And then shall it come to pass, that the spirits of the wicked, yea, who are evil—*for behold, they have no part nor portion of the Spirit of the Lord;* for behold, *they chose evil works rather than good;* therefore the spirit of the devil did enter into them, and take possession of their house—and these shall be cast out into outer darkness; there shall be weeping, and wailing, and gnashing of teeth, and this *because of their own iniquity,* being led captive by the will of the devil (Alma 40:13).

Surely it behooves us to make sure we never use our mouths (or any other parts of our bodies) in ways that would disqualify and exclude us from the eternal joys Heavenly Father has planned for us.

USING OUR MOUTHS TO SING

"I will sing unto the Lord as long as I live: I will sing praise to my God while I have my being" (Ps. 104:33).

Music has a power that puts mere words to shame. It can affect emotions and commitments in a way that words can never do alone. Singing can touch our hearts; it can reaffirm or even restore our beliefs and commitments; it can lift us and make us want to be better. That's one reason singing is a commanded part of our worship services. It not only nourishes our spirits, but serves as a form of instruction as we unite our voices in a unanimous expression of principle and value. "Let the word of Christ dwell in you richly in all wisdom; *teaching and admonishing one another in psalms and hymns and spiritual songs*, singing with grace in your hearts to the Lord" (Col. 3:16). Of course, Christ himself worshiped the Father in song. (See Matt. 26:30; Mark 14:26.)

Using our mouths to sing is also a form of *personal* worship. It's a therapeutic and healing outlet for the spiritual gratitude we feel as we become aware of God's influence in our lives. "The Lord is my strength and my shield; my heart trusted in him, and I am helped; therefore my heart greatly rejoiceth; and with my song will I praise him" (Ps. 28:7). "I will sing of thy power; yea, I will sing aloud of thy mercy in the morning: for thou hast been my defense and refuge in the day of my trouble" (Ps. 59:16).

The nourishment and healing we receive from expressing our feelings in song is so great that the scriptures also *command* us to express our personal worship and praise in song, "speaking to yourselves in psalms and hymns and spiritual songs, singing and making melody in your heart to the Lord" (Eph. 5:19). "Sing praises to God, sing praises: sing praises unto our King, sing praises" (Ps. 47:6).

Jesus said that he loves to hear us sing, "for my soul delighteth in the song of the heart" (D&C 25:12). He has given us this marvelous principle about personal worship through song: "Yea, *the song of the righteous is a prayer unto me,* and it shall be answered with a blessing upon their heads" (D&C 25:12). This verse teaches three principles. First, we can please our Savior and Heavenly Father by expressing our spiritual feelings in song. Second, the feeling of the heart matters to them more than the musicality of our own particular voice. And third, they receive that expression of feeling and worship as a prayer and promise to send blessings in response. Do you want a blessing? Sing praises! Sacred music has a marvelous power to slip past our discouragements and other mental blockades to nourish the heart. One of those promised blessings will be an increased resolve and power to say "no" to the temptations and discouragements that try to come between us and our God.

Elder Boyd K. Packer has advised that one effective way to deflect temptation is to sing or hum a hymn at the time of battle. This is an effective defense because our minds can focus on only one thing at a time. If we're experiencing the words and melody of a hymn, we can't be drooling over the pleasures of an enticing sin. Just think about it. How could we possibly give in to temptation while singing or humming a favorite hymn?

You don't need to memorize an entire song. Music is so powerful in affecting emotions that even if we can't remember all the words, just humming the melody will have an effect. Consider, for example, the power of just these two lines: "I need thee every hour, stay thou nearby; Temptations lose their power when thou art nigh" (*Hymns,* p. 98). Another powerful hymn is "How Firm a Foundation":

> Fear not, I am with thee, O be not dismayed,
> For I am thy God and will still give thee aid;
> I'll strengthen thee, help thee, and cause thee to stand,
> Upheld by my righteous, omnipotent hand. *(Hymns,* No. 85)

Many inspiring hymns are available to us in a variety of forms: sheet music, CDs, or radio broadcast, for example. If we select several that we like and then practice singing them when we're *not* discouraged or struggling with temptation, it will be natural to recall the words and melodies when we *are* beset by temptation. "I will sing praise unto thy name for ever, *that I may daily perform my vows*" (Ps. 61:8).

Music and the works of God are inseparable. The creation of the earth was celebrated with music as God's premortal children "sang together, and all the sons of God shouted for joy" (Job 38:7). The announcement of Christ's birth by a heavenly messenger was accompanied by a large heavenly choir described as "a multitude." (See Luke 1:13-14.) Perhaps some of us were in that choir.

One thing people comment about when they pierce the veil and behold the heavenly environment on the other side is the beautiful music enjoyed there. For example, when Lehi "saw God sitting upon his throne," he didn't describe the room or the throne itself, but reported the most compelling part of the vision—the beauty of what he heard as God was "surrounded with numberless concourses of angels in the attitude of singing and praising their God" (1 Ne. 1:8). During the three-day trance in which Alma received a conversion to Christ and a change of heart, he was also allowed to glimpse the musical wonders we'll someday enjoy in the kingdom of heaven. He, too, saw the enormous choir that provides music at the throne of God. He said the Lord was "*surrounded with numberless concourses* of angels, in the attitude of singing and praising their God" and that the singing he heard was so beautiful and uplifting that "my soul did long to be there" (Alma 36:22).

We can scarcely imagine the uplifting influence that *perfect* music, sung and performed by *perfected* musicians with *perfect* instruments, will have on us as we enjoy that perfection of expression throughout the eternities. But we can look forward to that time, knowing that music is an endless, unceasing part of the heavenly environment. Apparently the variety will be endless, too, for those who have seen the other side report that the heavenly choirs "sing *ceaseless* praises" (Morm. 7:7) and "they shall sing the song of the Lamb, *day and night* forever and ever" (D&C 133:56).

KEEPING OUR PROMISES

"My covenant will I not break, nor alter the thing that is gone out of my lips" (Ps. 89:34).

The most amazing experience I ever had with a person keeping her word occurred when I was taking calls at a dentist's office. Early one morning, before we opened, a lady called to cancel her husband's appointment for that day because he had died during the night! Would you have been concerned about such a detail in your time of grief?

That level of integrity certainly gave me a lot to think about. One of the attributes of God's perfection is his absolute commitment to deliver on every word he speaks and to keep every promise he makes, regardless of opposing circumstances, for "it is impossible for him to deny his word" (Alma 11:34) and "he executeth all his words" (2 Ne. 9:17). "For I will fulfill my promises which I have made unto the children of men" (2 Ne. 10:17). "For as I, the Lord God, liveth, even so my words cannot return void, for as they go forth out of my mouth they must be fulfilled" (Moses 4:30).

How often do we make excuses when someone confronts us for not keeping a promise? But the Lord makes no excuses. "Who am I, saith the Lord, that have promised and have not fulfilled?" (D&C 58:31). Following his encouraging example, we should be willing to say, "I have opened my mouth unto the Lord, *and I cannot go back*" (Judg. 11:35).

One of the principles that the scriptures stress in regard to the use of our mouths is the importance of keeping our word. If we promise something to the Lord, he should be able to count on us to do it. "That which is gone out of thy lips thou shalt keep and perform . . . according as thou hast vowed unto the Lord thy God, which thou hast promised with thy mouth" (Deut. 23:23). "If a man vow a vow unto the Lord, or swear an oath to bind his soul with a bond; *he shall not break his word*, he shall do according to all that proceedeth out of his mouth" (Num. 30:2).

Likewise, when we promise something to our family members or associates, they should also be able to count on us. "That which is gone out of thy lips thou shalt keep and perform" (Deut. 23:23). We've all heard the saying, "His word is his bond." Unfortunately, that ethic belongs more to a previous culture than it does to ours. We're always greatly disappointed when we count on someone to keep his promise, only to have him let us down.

Jesus told the interesting story of a father who needed help on a certain day from his two sons. The first son had plans and told the father he just couldn't help out that day, but later he rearranged his affairs and did go and help. The second son readily agreed to go, "and he answered and said, I go, sir: and went not" (Matt. 21:30). "Whether of them twain did the will of his father?" asked Jesus. "They say unto him, The first" (Matt. 21:31). "When thou shalt vow a vow unto the Lord thy God, *thou shalt not slack to pay it*: for the Lord thy God will surely require it of thee; and it would be sin in thee" (Deut. 23:21).

During a fierce battle in which the Lamanites were surrounded, defeated, and in peril of their lives, Moroni offered them their lives if they would simply take an oath to leave the country and never return in battle. These Lamanites, committed to the destruction of the Nephites, could easily have agreed to the oath, escaped, regrouped, and returned in spite of their promise. But in spite of their wickedness they had the integrity to say: "Behold, here are our weapons of war; we will deliver them unto you, but *we not suffer ourselves to take an oath unto you, which we know that we shall break*" (Alma 44:8).

Zoram, Laban's servant, was fooled by Nephi who, after slaying Laban, dressed in his clothes. Zoram was obedient to Nephi's instructions to take him to the brass plates until Laman and Lemuel became part of the group. Realizing there was treachery involved, Zoram panicked and was about to run away. Allowing Zoram to escape and sound the alarm would have thwarted their plans to obtain the scriptures. Urgently, Nephi testified to him of God's hand in their mission and invited him to flee with them to the new land God was leading them to.

In today's society, you'd expect a man in Zoram's outnumbered position to say anything to save his life. But in spite of Israel's apostasy, there were still high standards of ethics in keeping one's word. Nephi reported that "it came to pass that when Zoram had made an oath unto us, our fears did cease concerning him" (1 Ne. 4:37). Can people count on our word with that certainty? Can the Lord?

Scriptures emphasize that "when thou vowest a vow unto God, defer not to pay it," for it's better "that *thou shouldest not vow, than that thou shouldest vow and not pay*" (Eccl. 5:4-5). In our own time the Lord has reemphasized this principle: "And arise up and *be more careful henceforth in observing your vows*, which you have made and do make, and you shall be blessed with exceeding great blessings" (D&C 108:3). To attain the full image of Christ, we must learn to honor every word we speak—even when it requires sacrifice. "Verily I say unto you, all among them who know their hearts are honest, and are broken, and their spirits contrite, *and are willing to observe their covenants by sacrifice*—yea, every sacrifice which I, the Lord, shall command—they are accepted of me" (D&C 97:8). "Let the words of my mouth, and the meditation of my heart, be acceptable in thy sight, O Lord, my strength, and my redeemer" (Ps. 19:14).

Chapter Five

THE FACE

Our faces are the most expressive part of our anatomy, manifesting almost every emotion we feel. The face is so expressive of what's going on inside that we can often look at someone and discern exactly what that person is feeling—without a word being exchanged.

At my place of employment, I see approximately two hundred people a day. The things I see in their countenances never cease to amaze me. As people stand in line or approach my counter I see stress and anxiety, caused by an obvious pressure to get on to the next task. I see pain and sorrow. I see faces that are dull and numb. I see faces that are hard and weathered, both physically and emotionally. Occasionally I see a face that is serene and content. My duties there require me to check the photo identification of approximately fifty of these people each day. Looking at photo IDs takes one back in time, sometimes several years. The changes in appearance are interesting, but what is so captivating is that the vast majority of ID pictures I see show a different person than the one standing before me. Most of the time those photos show people who were happier, who were less hardened and stressed than they appear now. Every time I see the deterioration before me, I wonder what's happened to them in the last few years. This passing array of life-facts makes me wish for the opportunity to stop time and talk with these people to learn what lies in their past (and present) to cause such dramatic changes in their countenances.

A CHRISTLIKE COUNTENANCE

We know that mankind was created in the *physical* image of God. One of our challenges in this life is to so focus on the spiritual things of the kingdom that we not only become Christlike in our *behavior*, but that our faces, our very appearance or countenance, is transformed into his image. "And as we have borne the image of the earthy, we shall also bear the image of the heavenly" (1 Cor. 15:49).

To have a face that shines with the light of Christ should be the goal of every disciple. Every day should bring further effort to make our countenances into greater reflections of the light of Christ. This is but one of the many ways that we can develop our spiritual anatomy to become more like the Savior. When Alma asked, "Have ye received his image in your countenances?" he added several other questions by which you and I can look in a spiritual mirror to examine our own image. His questions were:

> "Can ye look up to God . . . with a pure heart and clean hands?"
> "Can you look up, having the image of God engraven upon your countenances?"
> "Have ye spiritually been born of God?"
> "Have ye experienced this mighty change in your hearts?" (See Alma 5:14, 19.)

One way we could characterize a Christlike countenance would be a face that manifests sensitivity and tenderness for others. Such a face displays warmth and kindness. Wickedness and rebellion tends to manifest in a face of coldness and even hardness, for "a wicked man hardeneth his face" (Prov. 21:29). "They have refused to receive correction: *they have made their faces harder than a rock*; they have refused to return" (Jer. 5:3).

While we can alter our outward appearance with cosmetics and hair care, we can't pretend our way into the image of Christ. That spiritually glowing appearance depends on the values we hold within. This visible transformation of our faces is a gradual progression from the face of the fallen, natural-man (or woman) to the face of a man or woman of Christ, as we grow and progress step by step. "But we all, with open face beholding as in a glass the glory of the Lord, are changed into the same image from glory to glory, even as by the Spirit of the Lord" (2 Cor. 3:18).

I want to share with you just how dramatic this spiritual change in a countenance can be. Part of my recovery from excommunication involved some counseling that I received from a wonderful LDS therapist. His spiritual guidance helped me obtain the change of heart and nature that are available to each of us through repentance and the atonement of Jesus Christ. Because we became such close friends, he later confided to me that on the day he first met me he recorded in his journal that I had the darkest face and countenance he had ever encountered. It was astonishing to him.

My parents were on a mission during my time of reconversion and spiritual rebirth. I hadn't realized what a *physical* change this inner birth had made in my face until my parents returned from their mission and came to visit. Coming from another town for their first visit in more than two years, they planned to meet with me at church on a Sunday and then we were to spend the day together. I waited in the foyer to greet them, but so great had been the change in my countenance that they actually walked right past me without even recognizing their own son! This experience taught me that our physical faces really do reflect *who* we are and *what* we are inside. As President Spencer W. Kimball stated:

> Not only does a person become what he *thinks*, but often he comes to *look* like it. If he worships the God of War, hard lines tend to develop on his countenance. If he worships the God of Lust, dissipation will mark his features. If he worships the God of Peace and Truth, serenity will crown his visage *(The Miracle of Forgiveness* [Salt Lake City: Bookcraft, 1969], p. 104).

You've seen this truth reflected in faces that are hard, the lines fixed, rigid, and stern from years of unpleasant life. You've seen faces that are soft, tender, and expressive. Perhaps you've seen evil, darkened faces like mine was from years of spiritual captivity and addiction. And perhaps you've seen beautiful, Christlike faces that glow with goodness, kindness, love, and concern for others.

Note: For an account of the author's excommunication and conversion to the Savior, see Steven Cramer, *The Worth of a Soul* [Springville, Utah: Cedar Fort Inc., 1983].

TURNING THE OTHER CHEEK

One of the most difficult parts of receiving Christ's image in our countenances is a special assignment Christ gave concerning our cheeks. "But I say unto you, that ye shall not resist evil, but whosoever shall smite thee on thy right cheek, turn to him the other also" (3 Ne. 12:39). Luke helps us understand the "resist not" part of Christ's instruction by telling us that we must not just "*turn* the other cheek," but "unto him that smiteth thee on the one cheek *offer* also the other" (Luke 6:29).

When Christ said that "ye shall not resist evil," he wasn't saying that we should not resist temptations and mortal desires to sin. He was teaching what he wanted us to do when we are the victims of unpleasant and unwanted circumstances such as criticism, disrespectful or offensive treatment, even deliberate ridicule—mental and emotional "slaps" upon the cheek, if you will. In these unjustified offenses we have a magnificent opportunity to rise above opposition and become more like Christ. Let's consider the example he set in this regard.

Several of the Old Testament prophecies concerning the Savior's atonement included specific, physical things that would be done to Christ's cheeks and the New Testament authors were careful to record their fulfillment. We're told that the torture inflicted on Jesus began late Thursday night after a mob of cutthroats, led by the traitor Judas, bound him and dragged him from the Garden of Gethsemane like a common criminal. The Bible tells us how, after the spiritual and physical agony he'd already suffered in the garden on our behalf, he was then beaten, taunted, ridiculed, and condemned by false witnesses throughout the rest of the night. They slapped his cheeks, they beat upon him with their fists; they ripped the hair of his beard from his cheeks; they even spit in his face. (See Matt. 26:67 and Luke 22:64.)

We need to recognize that this cruel, mocking abuse was not a foreordained, necessary part of the atonement for our sins, for that took place in the garden of Gethsemane and again on the cross. Slapping Jesus in the face was just a manifestation of the cruelty of these particular men. It was illegal treatment of a prisoner under Jewish law, and he could have justifiably protested. But rather than become involved in this debasing behavior (as we do when we rebel, resist, and protest), it's recorded that "[he] hid not [his] face from shame and spitting" (Isa. 50:6; 2 Ne. 7:6), but that "he [gave] his cheek to him that smiteth him" (Lam. 3:30).

When we receive physical or verbal slaps and put-downs, our normal response, our mortal, fallen, natural-man response, is to rebel, to resist, and even to slap back in kind, if not physically, then certainly verbally and emotionally. And it's *here*, at these times of deep and often unjustified hurts, that Christ is asking us not to resist, but to symbolically turn the face and offer our other cheek by allowing our attackers to continue doing or saying what they, in their own pain or immaturity, feel the need to do.

Even if you add together all the painful insults, all the cruel words that have cut your heart, all the ridicule, put-downs, and offensive things that have been said and done to you in your entire life, they can't begin to compare with the unjust, undeserved and irrelevant, unscheduled, unnecessary things done to Christ on that night. And yet, even during this night of physical and emotional exhaustion, even in this most vulnerable time, he "hid not his face from spitting," but willingly "*gave* his cheeks" to those who smote him. What a magnificent example Christ gave of the things we can do as we follow him toward a higher level of spirituality than those who go through life doing the "smiting"!

There are two major reasons for turning the other cheek. First, when we resist the offense and allow an unkind act to plunge us into feelings of resentment or self-pity, we lose spirituality and receptivity to the Holy Spirit. There's great power in predetermining our own position in life without responding, reacting to, or resisting the actions and offenses of others. But when we "offer the other cheek" by not resisting or retaliating, we gain in spiritual strength and become more Christlike.

Second, a passive yielding to such abuse may even touch the other person's conscience in a way that retaliation or protesting the injustice could never do. We see this principle illustrated in the death of the Anti-Nephi-Lehis, converted Lamanites who chose to die rather than defend themselves against the attacking Lamanites. Instead of responding to the attack with swords (resist not the evil), they "prostrated themselves before them to the earth, and began to call on the name of the Lord; and thus they were in this attitude when the Lamanites began to fall upon them, and began to slay them with the sword" (Alma 24:21). More than a thousand people died before the Lamanites realized what was happening. These evil, bitter, angry, and hardened warriors were so touched by this ultimate "turning of the other cheek" attitude that they threw down their swords and found their way into the church, bringing more new converts than the number of those that had perished.

> Now when the Lamanites saw that their brethren would not flee from the sword, neither would they turn aside to the right hand or to the left, but that they would lie down and perish, *and praised God even in the very act of perishing under the sword*—
> Now when the Lamanites saw this they did forbear from slaying them; *and there were many whose hearts had swollen in them* for those of their brethren who had fallen under the sword. . . . And it came to pass that they threw down their weapons of war, and they would not take them again, *for they were stung for the murders which they had committed* (Alma 24:23-25).

Discipline in my youth sometimes included an unexpected slap on the face. To me, the impact of this punishment was far worse than a few spanks on my bottom. It was not only more painful and immediate in its impact, but it was humiliating; it aroused powerful feelings of anger and resentment that I never felt when receiving a spanking. Perhaps you've experienced the difference, too. I think these kinds of feelings help us to understand why Christ chose the cheeks to symbolize the divine forgiving and yielding attitude he asks us to reach for at these moments. Truly the cheeks symbolize one of the greatest challenges in spiritual anatomy.

Note: The author does not want anyone involved in a serious spousal- or child-abuse situation to presume that the principle of "turning the other cheek" *in any way* implies that a victim of abuse should remain permanently in a harmful relationship. Any reader in such a situation is strongly encouraged to seek priesthood counsel. Elder H. Burke Peterson of the First Quorum of the Seventy wrote:

> The letters and phone calls the Brethren receive from faithful wives and children who are emotionally and physically abused in their own

homes continue to multiply. Their cries for help are heartrending. Their pleas and prayers are never-ending. It is tragic that too often husbands and fathers, even those who hold the priesthood, conduct themselves in their own homes in ways that would be censured in any other social setting. Countless heartaches and misshaped lives result from this unrighteous behavior *(Ensign,* July 1989, p. 7).

Anyone who knows or has cause to believe that a person has been or is a victim of physical, emotional, or sexual abuse has a solemn responsibility to do something constructive. There are many resources available in the community to assist in helping a family involved in domestic violence. Experts can be reached in various programs including shelters, victim assistance programs, and treatment providers. County attorneys' offices and law enforcement providers can assist in giving names, locations, and numbers for such programs. (In Utah, call 1-800-897-LINK [5465] for information and referral to domestic violence resources across that state.) Remember, domestic violence in the form of physical abuse is a crime.

JUDGING BY APPEARANCES

Perhaps it's because of this power of the face to communicate what we are and what we feel that so many of us are prone to judge each other by appearances, by the look of beauty (or lack of it) in physique and face—exactly what the Savior asked us not to do. *"Judge not according to the appearance,* but judge righteous judgment," he said (John 7:24). Throughout his mission, Paul made it clear that the Lord is not pleased with those who "glory in *appearance,* and not in heart" (2 Cor. 5:12).

Surely our opinions of each other would be more righteous if they were based on quality of discipleship rather than form and appearance. "Do ye look on things after the *outward* appearance?" Paul challenged (2 Cor. 10:7).

My wife's journal records her observations of a severely physically handicapped young man she met at a university and how he was accepted by his peers in spite of his appearance and other difficulties. She writes: "Though he can walk very haltingly, he suffers uncontrolled movements of his arms, hands and neck. His speech is forced, guttural, and difficult to understand. But his mind and intellect are sharp, as is evidenced by the very fact of his attendance at a university."

She further records: "It was marvelous to see how the young people accepted him into their midst. He was not treated one iota differently than any of the others. They talked and joked with him just as they did with each other. They were patient with his speech, waiting for his words to come out, not trying to help him along, but listening attentively and appreciatively. I could tell he felt no lack of self-worth. No pity, no special attention or help was offered him. It's a joy to see people who can see beyond a physical handicap to the person underneath and treat him accordingly."

When King Saul was rejected by the Lord, the prophet Samuel was sent to Bethlehem to choose a new king from the sons of Jesse. Seven of Jesse's eldest sons were presented to Samuel, and one of them really impressed him. "And it came to pass, when they were come, that he looked on Eliab, and said, Surely the Lord's anointed is before him" (1 Sam. 16:6). But the Lord had chosen David to be the next king, and said unto Samuel, "*Look not on his countenance*, or on the height of his stature; because I have refused him: *for the Lord seeth not as man seeth*; for man looketh on the outward appearance, but the Lord looketh on the heart" (1 Sam. 16:6-7). This should be great news for those of us who are not as beautiful or handsome as the standards of the world might make us wish to be.

GUILTY FACES

Have you ever been caught in a deliberate lie when someone challenged, "I don't believe you. Can you look me in the eye and say that?" At Halloween, we put on costumes and pretend to be someone else. Unfortunately, many of us do that all year long, hiding our real selves behind false faces and pretend selves.

A first-time offender named Robert stole a check. He was caught and arrested shortly after cashing it. Because of his lack of criminal experience, when he was asked for ID, he automatically presented his own driver's license and signed his own name on the back, complete with current address. Robert was not a very bright criminal. Nor are we very bright when we think we can hide who we really are inside by wearing pretend faces on the outside. "The shew of their countenance doth witness against them; and they declare [or reveal] their sins as Sodom, they hide it not" (Isa. 3:9).

There's an interesting story in the Book of Mormon about guilt showing on a face. A crime had been committed and the murderer was unknown to the people. Nephi, who was having difficulty being accepted as a prophet, told them who had committed the murder and how to confront the man so that he would confess. "From whence cometh this blood?" the investigators were to ask. "Do we not know that it is the blood of your brother?" (Hel. 9:32). Nephi then prepared them to recognize the signs of guilt in his face. "And then shall he tremble, and shall look pale, even as if death had come upon him. And then," Nephi instructed, "shall ye say: Because of this fear and this paleness which has come upon your face, behold, we know that thou are guilty" (Hel. 9:33). In fulfillment of Nephi's prophecy, the man guilty of the murder did deny his crime, but when challenged about the guilt showing in his face, he confessed and could no longer deny. (See Hel. 9:37-38.)

Ezra said, "O my God, I am ashamed and blush to lift up my face to thee, my God: for our iniquities are increased over our head, and our trespass is grown up unto the heavens" (Ezra 9:6). While we all have done things in our past that would cause our faces to blush in shame if they were known to those around us, we mustn't let these mistakes separate us from our Heavenly Father. He is more concerned about helping us remove the problems than he is about how they got there. Remembering the revelations our faces portray might help us to repent more quickly of the things we so often seek to hide behind false faces and countenances. "If iniquity be in thine hand, put it far away, and let not wickedness dwell in thy tabernacles. *For then shalt thou lift up thy face without spot*; yea, thou shalt be steadfast, and shall not fear" (Job 11:14-15).

How comforting it is to know that our mistakes and even our deliberate sins can all be erased and forgotten by the power of the atonement, literally blotted from the record as if they had never happened as we repent and trust Christ to cleanse us! And then we, like Enos, can look forward with joyful anticipation to seeing the face of God and allowing him to see and read our faces. "And I rejoice in the day when my mortal shall put on immortality, and shall stand before him; *then shall I see his face with pleasure*, and he will say unto me: come unto me, ye blessed, there is a place prepared for you in the mansions of my Father" (Enos 1:27).

Chapter Six
THE HANDS AND ARMS

"*. . . and mine elect shall long enjoy the work of their hands*" (Isa. 65:22).

Many of the blessings or chastisements we receive from the Lord are based specifically on the way we use our hands and arms. Considering that there are over 3,000 scriptures pertaining to the use of hands and arms, it's obvious the Lord is concerned about the way we use these precious extensions of our will. "Render unto them a recompense, O Lord, according to the work of their hands" (Lam. 3:64; also Prov. 12:14).

CHRIST'S HANDS AND ARMS

As we seek to develop our spiritual anatomy, we can know what God expects us to do with *our* hands and arms by recognizing what he does with *his*. For example, one of his highest goals is to "encircle [us] in the arms of [his] love" (D&C 6:20). How much happier and stronger our families would be if we never let a day go by when we did not hug and "encircle" someone in our family to express the love that otherwise might grow cold from being unexpressed!

"Behold," said Jesus, "*mine arm of mercy is extended* towards you, and whosoever will come, him will I receive; and blessed are those who come unto me" (3 Ne. 9:14). How could we expect the Savior's arms of mercy and forgiveness to welcome us in *our* repentance if we fail to extend our arms in mercy and acceptance to those who have disobeyed or wronged us? "And we see that his arm is extended to all people who will repent and believe on his name" (Alma 19:36). "Behold, he sendeth an invitation unto all men, *for the arms of mercy are extended towards them* and he saith: Repent, and I will receive you" (Alma 5:33).

I think of an encouraging parent whose hands and arms are extended invitingly to a child learning to walk—reaching, encouraging, offering shelter and welcome, just as Christ is offering his arms to us as we learn to repent and walk more obediently. The Savior's loving invitation is not occasional, but perpetual. "*All day long* I have stretched forth my hands unto a disobedient and gainsaying people" (Rom. 10:21). The Lord's patience is astonishing. He *never* gives up hoping we will come. To convince us of his patient love, he said, "I have spread out my hands all the day unto a rebellious people, which walketh in a way that was not good, after their own thoughts" (Isa. 65:2).

His generosity is not always appreciated. In fact, most of the time it's rejected, but that doesn't deter his yearning to help us. "*Notwithstanding I shall lengthen out mine arm unto them from day to day, they will deny me*; *nevertheless*, I will be merciful unto them, saith the Lord God, if they will repent and come unto me; *for mine arm is lengthened out all the day long*, saith the Lord God of Hosts" (2 Ne. 28:32). "And how merciful is our God unto us, for he remembereth the house of Israel, both roots and branches; *and he stretches forth his hands unto them all the day long*; and they are a stiffnecked and gainsaying people; but as many as will not harden their hearts shall be saved in the kingdom of God" (Jacob 6:4).

Our disobedience hurts the Lord because he has to watch us diminish our opportunities for blessings and joy. Yet through it all—even when we are so rebellious that it makes him angry—his hands are still stretched out to us in loving invitation to reconsider and return. "The Lord shall have no joy in their young men," said Isaiah, "for everyone is an hypocrite and an evil doer, and every mouth speaketh folly. *For all this his anger is not turned away, but his hand is stretched out still*" (Isa. 9:17; 2 Ne. 19:17). What a compelling example for the compassion we must extend with our own hands and arms!

"The eternal God is thy refuge," said Moses, "*and underneath are [his] everlasting arms*" (Deut. 33:27). How could we expect the Lord's arms to sustain *us* when we need help, if we don't eagerly offer the strength of our own hands and arms to comfort and assist *others* when we're in a position to do so? "Withhold not good from them to whom it is due, *when it is in the power of thine hand to do it*" (Prov. 3:27). "He shall feed his flock like a shepherd: *he shall gather the lambs with his arm*, and carry them in his bosom, and shall gently lead those that are with young" (Isa. 40:11).

We have many opportunities to use our arms to gather people and help them to feel welcome. I remember a time during my first year away from home when I felt unwelcome and unwanted in a new ward, where the people were unfamiliar and seemed unfriendly to me. Then, on what I had resolved would be my last Sunday there, a man I didn't know came beside me in the hall and, putting his arm around my shoulder as we walked, asked who I was and invited me to an activity. That one simple act of a disciple of Christ using his arms to "gather a lamb" helped me feel wanted and kept me attending. "O ye fair ones, how could ye have departed from the ways of the Lord! O ye fair ones, *how could ye have rejected that Jesus, who stood with open arms to receive you!*" (Morm. 6:17).

"How stupid of *them!*" we hastily judge. But how many of *us* are failing to use our hands and arms in appropriate ways that entitle us to receive Christ's love—ways that would create full fellowship with him? Glen L. Pace said:

> Our Lord and Savior Jesus Christ knows you intimately. He knows your name and he knows your pain. If you will approach your Father in Heaven with a broken heart and contrite spirit, you will find yourself miraculously lifted into the loving and comforting arms of the Savior *(Ensign,* Nov. 1987, p. 41).

As disciples of Christ, let us not only find our own way into those arms of love, but make an effort to use our own hands and arms in ways that can lead others there as well.

CRUEL HANDS ABUSE THE SAVIOR

Throughout his mission, the Savior taught his disciples that part of his mission was to "be *delivered into the hands* of sinful men" (Luke 24:7; 9:44; Mark 9:31). When the mob arrived in the Garden of Gethsemane to arrest him, he said, "The hour is come; behold, the Son of man is betrayed into the hands of sinners" (Mark 14:41). "Then came they, and laid hands on Jesus, and took him" (Matt. 26:50).

Throughout the night they took turns taunting him and beating him. "Then did they spit in his face, and buffeted him; and others *smote him with the palms of their hands*" (Matt. 26:67). Jesus didn't respond to most of their questions and taunts, but when the high priest asked him of his doctrine he did answer. Of course, they didn't like his answer. "And when he had thus spoken, one of the officers which stood by struck Jesus with the palm of his hand, saying, Answerest thou the high priest so?" (John 18:22). This unwarranted torture and ridicule was not a foreordained, essential part of the planned atonement. It was just rampant, vicious cruelty, as they mocked him, saying, "Hail, King of the Jews! *and they smote him with their hands*" (John 19:3).

It's easy for us to condemn that cruel use of hands and mouths. We can smugly say that we would never use *our* hands to betray the Savior, or to show unkindness to him. But is that true? For example, isn't it an act of betrayal to use our hands to slap a child in anger? "Inasmuch as ye have done it unto one of the least of these my brethren, ye have done it unto me" (Matt. 25:40). The use of our hands or arms to do anything that would make us ashamed for him to know is a form of betrayal. The deliberate use of our hands to masturbate, for example, or to commit other sexual sins is also a mockery of his sacrifice and atonement.

Christ gave his hands and feet to the Roman soldiers, willingly allowing them to drive spikes through his flesh as they nailed him to the cross. He allowed it that he might die for us in the process of conquering spiritual and physical death. These were the hands and feet of the creator of the universe! Now he asks us not to take that selfless sacrifice for granted, but to remember it; to ponder and appreciate it; to be moved by it. "Behold the wounds which pierced my side," he pleads, "and also the prints of the nails in my hands and feet" (D&C 6:37). As we contemplate and mentally "behold" and appreciate those wounds each Sunday during the sacrament service, there comes a power of determination to honor his pain with new devotion and more obedient hands.

I once worked for a business that required delivery people to record the starting and ending mileage of their vehicles in our computers each day. I was often amused to see them write the mileage on the palms of their hands so they could remember to record it when they got inside to the computer.

But then I discovered a scripture that sobered me in the way it refers to the nail prints in the hands of Jesus. "Behold, I have graven thee upon the palms of my hands" (Isa. 49:16). This scripture moves me because I've learned that the pain and sacrifice Jesus suffered in his atonement was personal, not only *for* each individual person, but also *because* of each person. Through sacred experiences I've learned that the pain and anguish he suffered was greater because my sins were included. I believe the agony he suffered, while infinite for *all* of mankind, was also *individual*, for *each* of mankind. "Behold, I have graven thee upon the palms of my hands" (Isa. 49:16).

I believe that when he thinks of his agony in the Garden of Gethsemane and again on the cross, he knows exactly what *my* sins cost him in suffering. So when he looks at those scars in his hands, he remembers *me* and what *my* salvation (or the possibility of my salvation) meant in his agony. This awareness makes me want to try harder in my daily life to live and use my hands in a way that will remember and honor what he did for me—and because of me.

HAVING CLEAN HANDS OR WICKED HANDS

"They shall believe in Christ, and worship the Father in his name, *with pure hearts and clean hands*" (2 Ne. 25:16).

One of the standards required of those who are acceptable to the Lord and allowed to stand in his holy places is to have "clean hands, and a pure heart" (Ps. 24:4). Thus Alma challenged, "Can ye look up to God at that day with a pure heart and clean hands?" (Alma 5:19). The scriptures often equate sin with *unclean* hands. Thus we are challenged to "entangle not yourselves in sin, *but let your hands be clean, until the Lord comes*" (D&C 88:86). "And evil will befall you in the latter days; because ye will do evil in the sight of the Lord, *to provoke him to anger through the work of your hands*" (Deut. 31:29).

The Lord has asked "that *every man should take righteousness in his hands* and faithfulness upon his loins" (D&C 63:37). But sexual immorality is the major sin of our times, and it almost always involves the use of our hands. The moral responsibility to keep our hands from abusing our bodies or committing sexual sin may be one thing Isaiah had in mind when he said, "Blessed is the man that . . . keepeth his hand from doing any evil" (Isa. 56:2). President Spencer W. Kimball gave these sobering words about the misuse of our hands:

> Prophets anciently and today condemn masturbation. It induces feelings of guilt and shame. It is detrimental to spirituality. *It indicates slavery to the flesh, not that mastery of it and the growth toward godhood which is the object of our mortal life.* Our modern prophet has indicated that no young man should be called on a mission who is not free from this practice (The Miracle of Forgiveness [Salt Lake City: Bookcraft, 1969], p. 78).

> Woe to them that devise iniquity, and work evil upon their beds! When the morning is light, they practice it, *because it is in the power of their hand* (Micah 2:1).

Like every other part of the body, our fingers can also be used for righteousness or wickedness. We can use them, for example, to call a friend who needs a word of encouragement, or to dial for some mental pollution on a 900 sex number or computer keyboard.

One of the more frequent sins that we commit with our fingers is using them to criticize. A pointing finger is associated with scorn and ridicule, as with those in Lehi's dream of the Tree of Life: "And they were in the attitude of mocking and pointing their fingers towards those who had come at and were partaking of the fruit" (1 Ne. 8:27). The Lord has warned us of the accountability we take upon ourselves when we point our finger at others in judgment: "Judge not, that ye be not judged. *For with what judgment ye judge, ye shall be judged:* and with what measure ye mete, it shall be measured to you again" (Matt. 7:1-2).

Paul said, "Thou art inexcusable, O man, whosoever thou art that judgest: *for wherein thou judgest another, thou condemnest thyself;* for thou that judgest doest the same things" (Rom. 2:1). If that condemnation seems harsh, be aware that Jesus has repeated it in our own day, saying, "Those who cry transgressions do it because they are the servants of sin, and are the children of disobedience themselves" (D&C 121:17).

It's interesting to note what Christ did when asked to judge a woman taken in the very act of adultery. Refusing to join the self-righteous mob who were shouting and pointing their fingers at her accusingly, he "stooped down, and with his finger wrote on the ground, as though he heard them not" (John 8:6). If Christ were not a God of love, forgiveness, and acceptance, we might have expected him to point his own finger at her accusingly. "So, you've filthied yourself with the greatest sin there is next to murder! How in the world did you ever fall to this level? Are you sorry? Have you learned your lesson?" But those were not his words to her, nor is this ever his attitude toward one who comes to him in repentance. When he arose and found the crowd gone, he said, "Woman, where are those thine accusers? hath no man condemned thee? She said, No man, Lord. And Jesus said unto her, *Neither do I condemn thee*: go, and sin no more" (John 8:10-11). Since "God sent not his Son into the world to condemn the world" (John 3:17), it seems obvious that we should be careful about pointing our own fingers in judgment or criticism.

As with all consequences of the use of our agency, there are blessings for righteousness and tutoring punishments for disobedience. "Woe unto the wicked! it shall be ill with him: for *the reward of his hands shall be given him*" (Isa. 3:11; also 2 Ne. 13:11). "And I will recompense them according to their deeds, and according *to the works of their own hands*" (Jer. 25:14).

So serious and long-lasting are the consequences of the sins we commit with our hands that the Savior suggested a remedy that we could apply to judging, masturbation, or other sins related to sexual immorality: "And if thy hand offend thee, cut it off: it is better for thee to enter into life maimed, than having two hands to go into hell, into the fire that never shall be quenched" (Mark 9:43).

It isn't likely that Christ really intended for us to grab a knife and start hacking off a limb. Rather, he was showing the importance of controlling both our thoughts and the actions of our hands as they carry out our unclean thoughts. "If iniquity be in thine hand, put it far away, and let not wickedness dwell in thy tabernacles" (Job 11:14).

I believe Pilate saw goodness in Christ. According to Peter, if Pilate could have found a political solution that would spare Jesus, he would have released him. But that was not to be. (See Acts 3:13.) So, "when Pilate saw that he could prevail nothing, but that rather a tumult was made, he took water, *and washed his hands* before the multitude, saying, I am innocent of the blood of this just person: see ye to it" (Matt. 27:24).

Hands, like the rest of our anatomy, must be cleansed of bad habits. But sinful acts committed by hands are not so easily erased. We'll be held accountable for every unworthy act that we perform with our hands and arms unless we wash them symbolically in the blood of Christ. When we do make mistakes, the invitation is, "Draw nigh to God, and he will draw nigh to you. Cleanse your hands, ye sinners; and purify your hearts, ye double minded" (James 4:8). By receiving the cleansing influence of Christ's atonement, we're able to "draw nigh" to God and cleanse our hands of bad habits and unworthy acts through repentance. "Prepare yourselves, and sanctify yourselves; yea, purify your hearts, and *cleanse your hands and your feet before me that I may make you clean*" (D&C 88:74).

Each moral action we take with our hands makes us stronger or weaker. Every time we do something with our hands that the Savior would not approve of, we grow weaker and more susceptible to making the same mistake each time the temptation is presented. Each time we choose the right and use our hands for righteousness, we develop strength and make it easier to choose correctly when the temptations come. "*And he that hath clean hands shall be stronger and stronger*" (Job 17:9). "The Lord rewarded me according to my righteousness: *according to the cleanness of my hands hath he recompensed me*" (2 Sam. 22:21; also Ps. 18:20, 24).

THE POWER OF CHRIST'S HANDS

"Know ye not that ye are in the hands of God?" (Morm. 5:23).

When God called Moses to deliver the children of Israel from Egyptian bondage, he said, "I have heard the groaning of the children of Israel, whom the Egyptians keep in bondage; and I have remembered my covenant" (Ex. 6:5). The Lord realized there would be skepticism and doubt when Moses announced himself as their deliverer, so he instructed Moses to make this promise to them: "Wherefore say unto the children of Israel, I am the Lord, and I will bring you out from under the burdens of the Egyptians, and I will rid you out of their bondage, *and I will redeem you with a stretched out arm* and with great judgments" (Ex. 6:6).

For the next thousand years, Israel's prophets would bear testimony of the miracles and wonders that God's "stretched out arm" had performed on their behalf. "And the Lord brought us forth out of Egypt with a mighty hand, and with an outstretched arm, and with great terribleness, and with signs, and with wonders" (Deut. 26:8). One of the things the Lord has wanted us to understand *today* is the continuing unlimited power of his hands and arms to do whatever is required on our behalf. "For I am God, and *mine arm is not shortened*; and I will show miracles, signs, and wonders, unto all those who believe on my name" (D&C 35:8).

As we come to God in prayer, the power of his hands to affect our lives is never limited by difficult circumstances. "His purposes fail not, *neither are there any who can stay his hand*" (D&C 76:3). Today, in our modern times with many complex cultural limitations, the Lord has promised, "No power shall stay my hand" (D&C 38:33), "and my arm is stretched out in the last days, to save my people Israel" (D&C 136:22). "O house of Israel, is my hand shortened at all that it cannot redeem, or have I no power to deliver?" (2 Ne. 7:2). "For ye are the children of Israel, and of the seed of Abraham, and ye must needs be led out of bondage by power, and with a stretched out arm" (D&C 103:17).

TRUSTING THE ARM OF FLESH

Choosing to rely on our own strengths, or the power and resources of other men, in preference to God's help is known in the scriptures as "relying on the arm of flesh." The Lord has made it plain that those who so choose will be cursed by their choice because no *man* can cleanse us from our sins or make us worthy to return to Heavenly Father, no matter how skilled he is or righteous his intent. "Thus saith the Lord; Cursed be the man that trusteth in man, and maketh flesh his arm, and whose heart departeth from the Lord" (Jer. 17:5). "Cursed is he that putteth his trust in man, or maketh flesh his arm, or shall hearken unto the precepts of men, save their precepts shall be given by the power of the Holy Ghost" (2 Ne. 28:31).

The anti-Christ, Korihor, taught that reliance on Jesus Christ "is the effect of a frenzied mind; and that derangement of your minds," he said, "comes because of the traditions of your fathers, which lead you away into a belief of things which are not so" (Alma 30:16). But as we mature spiritually, we learn that to achieve true victory over our fallen natures, our *self*-centeredness and total reliance upon the ways of man ("arm of the flesh") must be surrendered to a *Christ*-centeredness and reliance on him. He has told us that he is our *only* path back to the Father. (See John 14:6.) While we use all the wisdom, strength, and ability with which we've been blessed, we still must learn to rely on Christ for the grace and divine assistance that will enable us to live the life he desires of us.

If any Church leader has ever taught about hard work, independence, and self-reliance, it was the great colonizer, Brigham Young. And yet he referred to this process of surrender and change of foundation from the arm of flesh to Christ as "the greatest and most important of all requirements."

> The greatest and most important of all requirements of our Father in Heaven and of his Son Jesus Christ . . . is to believe in Jesus Christ, confess him, seek to know him, cling to him, make friends with him. Take a course to open and keep open a communication with your elder brother or file leader—our Savior *(Journal of Discourses,* Vol. 8, p. 339, as quoted in the Relief Society Manual, 1982, p. 24).

> O Lord, I have trusted in thee, and I will trust in thee forever. I will not put my trust in the arm of flesh; for I know that cursed is he that putteth his trust in the arm of flesh. Yea, cursed is he that putteth his trust in man or maketh flesh his arm (2 Ne. 4:34).

As the "natural man" strives to overcome his weaknesses and faults, it's normal for him to focus his thoughts and emotions, his confidence and reliance on himself. After all, ever since we were children we've been waiting until we were grown up and could do it by ourselves. This introduces us to a fine line between efforts that are appropriate and well-intended, and the misleading "I-must-do-it-all-by-myself" attitude that can become self-defeating.

Elder Marion D. Hanks once told of a young boy who was assigned by his father to remove a large rock from the yard. The little boy accepted the difficult assignment confidently. After all, if his father hadn't believed in him, why would he assign the task? He tugged and pushed and lifted and struggled, but all to no avail. Even when he enlisted the help of his neighborhood friends, the boulder would not budge. Reluctantly he reported to his father that he could not move the rock. "Have you done all you could?" asked the father. "Yes," said the boy. "Are you sure you have tried *everything*?" the father persisted. "Yes, I've tried everything." "No, son, you haven't tried everything. You haven't asked *me* for help." Then Elder Hanks said:

> Some of us may be less happy than we could be because of arrogance or pride. We think we are sufficient unto ourselves. We think we do not need God or his Christ. Why do so many of us "heirs of God, joint-heirs with Christ," fail to go to him? He is anxious to help. But he wants us to learn our need for him *(Ensign,* July 1972, p. 105).

Of course, the Lord expects us to use our own strengths and abilities to do as much for ourselves and others as we can, for "it is by grace that we are saved, after all we can do" (2 Ne. 25:23). But it's also imperative that we recognize that no one can ever do enough, by his efforts alone, to live the obedient life that will result in exaltation. If we could, why would we need a Savior? As we eagerly do all that we can, let us not offend the Savior by forgetting the grace and enabling power he is anxious to add to our efforts. Consider the following words from the 1969-70 Gospel Doctrine Manual:

> "I am the Way," the Savior tells those who hope to find the means, the way to create heaven. *Only in him can any man find the strength, the power and ability to live a godly life*. Only in Christ is there power to transform the human mind and the human heart . . . . Only in Jesus Christ can any man learn the truth of what he is and how he can be changed from what he is to do the good for which he hopes *(In His Footsteps Today* [Salt Lake City: Deseret Sunday School Union, 1969], p. 4).

BECOMING INSTRUMENTS IN HIS HANDS

Throughout his mortal ministry, Christ used *his* hands to show us what we should be doing with *ours*. Constantly he reached out to touch, heal, and comfort others. "And he laid his hands on every one of them, and healed them" (Luke 4:40). "And Jesus, moved with compassion, put forth his hand, and touched him" (Mark 1:41). When they brought little children to him, "he took them up in his arms, put his hands upon them, and blessed them" (Mark 10:16).

Our hands may not heal *physical* infirmities with every touch as his did, but there is great healing power in the loving, kind touch of a hand. We can learn to reach out with our hands in loving and compassionate ways and touch the lives of others to heal spiritual and emotional wounds and reflect the love of Christ, whom we are trying to serve and emulate.

Just as the Savior spent his mortal ministry using his hands and arms to touch and love his brothers and sisters, he is now seeking to continue that ministration through us. "*And their arm shall be my arm . . . and they shall fight manfully for me*" (D&C 35:14). To those who are willing to serve him, he says, "*I will make an instrument of thee in my hands* unto the salvation of many souls" (Alma 17:11) and "*by your hands I will work a marvelous work* among the children of men" (D&C 18:44). "And this is the blessing which hath been bestowed upon us, that we have been made instruments in the hands of God to bring about this great work" (Alma 26:3). "And it came to pass that Alma, being a man of God, being exercised with much faith, cried, saying: O Lord, have mercy and spare my life, *that I may be an instrument in thy hands* to save and preserve this people" (Alma 2:30).

Peter's faith enabled him to climb out of his boat and walk upon the water when Christ invited him to do so. "But when he saw the wind boisterous, he was afraid; and beginning to sink, he cried, saying Lord, save me" (Matt. 14:30). There's a great lesson about the use of *our* hands in what happened next. "And *immediately* Jesus stretched forth *his* hand, and caught him" (Matt. 14:31). How quick are we to use our hands to serve? Are we alert and poised to "reach out immediately"? Or are our hands selfish and sluggish? Do we need motivation (or feelings of obligation) before we grudgingly move into action?

The Lord has asked us to develop *valiant* hands. "Whatsoever thy hand findeth to do, do it with thy might" (Eccl. 9:10) and "Be ye strong therefore, and let not your hands be weak: for your work shall be rewarded" (2 Chr. 15:7). We're also told to develop *unselfish* hands, hands that are ready and eager to serve willingly: "Thou shalt not harden thine heart, nor shut thine hand from thy poor brother" (Deut. 15:7).

I'm fascinated by an event that occurred at the time of Christ's arrest. Peter, not understanding the divine purpose of the Savior giving himself into the power of wicked men, "stretched out his hand, and drew his sword, and struck a servant of the high priest's, and smote off his ear" (Matt. 26:51). One person who might not have deserved the healing power of Jesus was surely this man who came to arrest Christ. But one of the majestic characteristics of our Savior's love is that it's never based on our *merit*, but rather on the simple fact that we are his brothers and sisters, regardless of what we've done. So Jesus said to the angry crowd, intent on dragging him away: "Suffer ye thus far. And he touched his ear, and healed him" (Luke 22:51). It's easy to love and serve those who "deserve" it, but do we have the ability to use our hands in the service of those who continually offend or abuse us? Surely it's a worthy goal to reach for.

Consider the inspiring words of this song:

His hands would serve His whole life through,
Showing men what hands might do. Giving, ever giving endlessly.

Each day was filled with selflessness,
And I'll not rest till I make of my hands what they could be,
Till these hands become like those from Galilee.

They take His hands—His mighty hands—those gentle hands,
And then they pierce them—they pierce them,
He lets them because of love!
From birth to death was selflessness,
And clearly now I see Him with His hands calling to me.

And though I'm not yet as I would be,
He has shown me how I could be.
I will make my hands like those from Galilee.

("His Hands," written by Kenneth Cope. 1989 Mohrgud Music (BMI), a subsidiary of Excell Entertainment Group, Inc. From the album *Greater Than Us All.* Used by permission.)

It is recorded that "God wrought special miracles by the hands of Paul" (Acts 19:11). We have been commanded to encourage each other, to "lift up the hands which hang down" (D&C 81:5), and to "strengthen ye the weak hands" (Isa. 35:3). Will the record of our lives report that God was able, or unable, to use our hands when he needed them?

Chapter Seven

THE FEET

"He knoweth thy walking. . . ." (Deut. 2:7).

Shortly after we were married, my wife and I were stationed in Duluth, Minnesota. Downtown Duluth, on the shore of Lake Superior, is built on rather steep hills, and each block is considerably higher than the one below. One day we parked on First Street and stepped onto the sidewalk to go down the steep hill to Main Street. During the winter, the sidewalks are often covered with clear ice, formed by a light sleet rain. The weather on this day was clear and we didn't realize the sidewalk was covered with the slippery ice until we instantly began sliding downward at a rapidly increasing speed.

We grabbed each other by the arm and managed to maintain our balance as our sliding speed continued to increase. We hardly had time to congratulate ourselves about sliding on our feet instead of our bottoms before we realized the danger in shooting out into traffic when we reached the bottom of the hill. Fortunately, I was able to hook one arm around a light pole, which swung us in a circle and broke our speed with no harm done. Laughingly, we went on our way, excited by the unexpected adventure.

Unknowingly, our feet had suddenly placed us in physical peril. Today the world is filled with spiritual perils, and unless we learn to control where our feet take us, we can be led into paths that are dangerous to our eternal destiny. As we go through mortality, we spend countless hours walking on our feet. A significant part of mortal probation includes the testing of our feet, for "through them I may prove Israel," said the Lord, "whether they will keep the way of the Lord to *walk* therein" (Judg. 2:22). We, the eternal spirit entities inside the bodies doing the walking, are the ones in charge of those feet. We are the ones responsible for the paths they take and the consequences of those directional choices.

There are almost 2,000 revelations about the ways we should and should not use our feet! The Lord has asked his disciples to use their feet for righteousness. He asked us, for example—

- To "walk in the meekness of my Spirit" (D&C 19:23).
- To "walk honestly" (Rom. 13:13).
- To "walk humbly with thy God" (Micah 6:8).
- To "walk before me in truth with all [your] heart and with all [your] soul" (1 Kgs. 2:4).
- And to go through life "walking in all holiness before me" (D&C 21:4).

As we go through each day of mortal probation, the Lord is watching to see how we use our feet, "for his eyes are upon the ways of man, and he seeth all his goings" (Job 34:21).

RIGHTEOUS FEET AND WICKED FEET

Did you realize that we usually need our feet to carry out our lusts? Peter warned that "there shall come in the last days scoffers, *walking* after their own lusts" (2 Pet. 3:3). Paul said that "there should be mockers in the last time, who should walk after their own ungodly lusts" (Jude 1:18). Some examples might be walking into a video or magazine store to rent or purchase something inappropriate; walking into a theater to view an inappropriate movie; or using our feet to go on a date where we violate the laws of chastity. Many sins of immorality require the use of our feet.

In contrast, the Lord has commanded us to "*walk* in the paths of virtue before me" (D&C 25:2). Using our feet to attend church or the temple would be walking in virtue, as would using our feet to do home teaching or visiting teaching, or serving a person in need. The endless miles walked on a mission is another example of walking in virtue. Paul challenged us to "walk worthy of the vocation wherewith ye are called" (Eph. 4:1). In the Church we have many callings or "vocations," but as disciples of Christ, our primary "vocation" or spiritual assignment is to follow the example of Christ as closely as we can in all things and to "walk worthy of God, who hath called you unto his kingdom and glory" (1 Thes. 2:12). It's obvious that we'll never inherit his kingdom or his glory unless we train our feet to walk as he walked, so that we may do and experience the things he has planned for our earth schooling.

Alma challenged his congregation with this question: "Have ye walked, *keeping yourselves blameless* before God?" (Alma 5:27). King Benjamin explained that his teachings were intended to help his people to "*walk guiltless* before God" (Mosiah 4:26). Do you believe that mortal feet can walk "blameless" and "guiltless"? Perhaps not—at least not alone. But with Christ's help, we can do everything he expects of us. The issue with our feet is not our *ability* to walk as the Lord expects, but our desire and *willingness* to do so. It's natural to doubt our ability to "walk worthy of the Lord unto all pleasing" (Col. 1:10) because, by ourselves, none of us could walk such a perfect path. But God doesn't expect that. Since "the Lord giveth no commandments unto the children of men, save he shall prepare a way for them that they may accomplish the thing which he commandeth them" (1 Ne. 3:7), he's looking for those *willing* to walk in his steps. He walks with those who are willing to receive him, bestowing the grace and power that enables them to attain what they choose or will to do.

We all falter and make mistakes as we walk in mortality. But repentance and the cleansing atonement of Jesus Christ can cleanse and remove those mistakes so that when our walking is done, we stand clean, blameless, and guiltless. If we decide to take charge of our feet and guide them in the Lord's path, the time will surely come when we can say with the Psalmist, "Judge me, O Lord; for I have walked in mine integrity" (Ps. 26:1), and with Hezekiah, "I beseech thee, O Lord, remember now how I have walked before thee in truth and with a perfect heart" (2 Kgs. 20:3).

USING OUR FEET TO TRAMPLE

I once heard a talk in which the speaker suggested that one way to motivate ourselves to keep our covenants is to imagine that each time we choose to deliberately sin, that act of disobedience is like taking the hammer from the Roman guard and saying, "Let me pound the spikes into the Savior's hands and feet myself." Or, as Paul described the rebellious who would not accept repentance, "seeing they crucify to themselves the Son of God afresh, and put him to open shame" (Heb. 6:6; also D&C 76:35). In a similar way, the scriptures have compared our indifference to the Lord's blessings and commandments as the act of using our feet to stomp on him, to symbolically trample him under our feet. "For the things which some men esteem to be of great worth, both to the body and soul, others set at naught and trample under their feet" (1 Ne. 19:7).

We who have guilty consciences may admit that we might trample a *commandment* we don't like or don't keep very well, but surely we would never do such a thing to the Savior himself. However, Nephi explains in the same verse that a lukewarm attitude of indifference to the Savior is the same as deliberate personal mocking, ridicule, and rejection: "Yea, even the very God of Israel do men trample under their feet; I say, trample under their feet but I would speak in other words—*they set him at naught,* and hearken not to the voice of his counsels" (1 Ne. 19:7).

Paul spoke of the enormity of the sin of indifference and taking the Savior's love and atonement for granted when he said: "Of how much sorer punishment, suppose ye, shall he be thought worthy, who hath trodden under foot the Son of God, and hath counted [or regarded] the blood of the covenant, wherewith he was sanctified, an unholy [unimportant] thing, and hath done despite unto the Spirit of grace?" (Heb. 10:29). As Alma cried to his audience: "And now my beloved brethren, I say unto you, can ye withstand these sayings; yea, *can ye lay aside these things, and trample the Holy One under your feet?*" (Alma 5:53). "Reproach hath broken my heart; and I am full of heaviness: and I looked for some to take pity, but there was none; and for comforters, but I found none" (Ps. 69:20).

The act of *deliberate disobedience* is also labeled by scripture as the act of using our feet to "trample" upon the commandments, as if we were trying to stomp them into the ground so that we can't see them or be obligated by them anymore. "Ye know that ye do transgress the laws of God, *and ye do know that ye do trample them under your feet*"

(Alma 60:33). "They had become exceedingly wicked; yea, the more part of them had turned out of the way of righteousness, *and did trample under their feet the commandments of God*" (Hel. 6:31).

The more we know and love the Savior, the deeper our devotion and loyalty grows and the harder it becomes to disobey him. So the far greater use of our feet is to "stand in awe, and sin not" (Ps. 4:4). "Let all the earth fear the Lord: let all the inhabitants of the world stand in awe of him" (Ps. 33:8). For me, one of our sacrament hymns expresses that awe:

I stand all amazed at the love Jesus offers me . . .
I tremble to know that for *me* he was crucified, that for me, a sinner, he suffered, he bled and died.
I marvel that he would descend from his throne divine To rescue a soul so rebellious and proud as mine . . . *(Hymns,* No. 193.)

WALKING IN STRAIGHT PATHS OR CROOKED

"Thou tellest my wanderings. . . ." (Ps. 56:8).

One of mortal man's challenges is to focus on the purpose of this earth life and to train our feet to pursue a *straight* and undeviating course toward our eternal destination, instead of wandering through the many distractions of life. "Make straight paths for your feet," counseled Paul (Heb. 12:13) and "walk in his paths, which are straight," taught Alma (Alma 7:9). "Behold, the way for man is narrow, but it lieth in a straight course before him" (2 Ne. 9:41).

Because we live in a fallen state, one of our handicaps is undependability. We mean well, but we waver and vacillate. Our mortal feet wander in crooked paths even though we've often been challenged to "enter into the narrow gate, and *walk in the strait path* which leads to life" (2 Ne. 33:9). Our wandering feet are not necessarily wicked. It's just that there are so many interesting places to go and time-consuming things to do that it's sometimes difficult to prioritize and keep our feet walking on an undeviating course toward eternal goals. For this reason, we're admonished to "ponder the path of thy feet" (Prov. 4:26). "But behold, for none of these can I hope except they shall be reconciled unto Christ, and enter into the narrow gate, *and walk in the strait path which leads to life,* and continue in the path until the end of the day of probation" (2 Ne. 33:9).

So we have choices. We can use our feet to walk the straight path with determination and persistence, or we can allow them to wander, robbing us of precious time and progression. "*They have loved to wander,*" said the Lord of ancient Israel, "*they have not refrained their feet,* therefore the Lord doth not accept them: he will now remember their iniquity, and visit their sins" (Jer. 14:10). "With my whole heart have I sought thee: *O let me not wander* from thy commandments" (Ps. 119:10).

We're told that those who insist on walking in crooked paths "shall not know peace" (Isa. 59:8). God, on the other hand, "doth not walk in crooked paths, neither doth he turn to the right hand nor to the left, neither doth he vary from that which he hath said, therefore his paths are straight" (D&C 3:2). As we train our feet to follow our Savior, we too should be seeking a straight and undeviating course that refuses to wander. "For behold, it is as easy to give heed to the word of Christ, *which will point to you a straight course to eternal bliss,* as it was for our fathers to give heed to this compass, which would point unto them a straight course to the promised land" (Alma 37:44).

If we want to attain our goals, we must train our feet to press *forward* and not allow them to wander or go backwards. The Lord's invitation to ancient Israel was the same as it is to us today: "Obey my voice, and I will be your God, and ye shall be my people. . . . But they hearkened not, nor inclined their ear, but walked in the counsels and in the imagination of their evil heart, *and went backward, and not forward*" (Jer. 7:23-24). These unwise people forfeited the blessings the Lord was trying to give them. Will we? "Brethren, shall we not go on in so great a cause? *Go forward and not backward.* Courage, brethren; and on, on to the victory!" (D&C 128:22).

> Wherefore, *ye must press forward* with a steadfastness in Christ, having a perfect brightness of hope, and a love of God and of all men. Wherefore, if ye shall press forward, feasting upon the word of Christ, and endure to the end, behold, thus saith the Father: Ye shall have eternal life (2 Ne. 31:20).

Apparently the consequences of the path our feet take are so important that Christ said: "If thy foot offend thee, cut it off: [for] it is better for thee to enter halt into life, than having two feet to be cast into hell, into the fire that never shall be quenched" (Mark 9:45). In today's world we wouldn't interpret this counsel literally, but surely we could "ponder the path of our feet" and cut them off from the freedom to wander and lead us to evil. Surely we could join Nephi in his personal prayer for God to "make my path straight before me" and to help him "walk in the path . . . [and] . . . be strict in the plain road." (See 2 Ne. 4:32-33.)

USING OUR FEET TO WALK IN LIGHT OR DARKNESS

I remember a frightening experience as a child, awakening at night and not being able to find the door. My mother had to rescue me from a corner, where I stood crying helplessly. Have you ever had the experience of trying to walk in a place of total darkness? Such an experience makes one aware, very quickly, how easy it is to take the light for granted. Wouldn't it be stupid to stumble around in the dark when it takes only the flick of a switch to turn the lights on? And yet, the majority of mankind stumble through this mortal life in spiritual darkness because they will not come to Christ or walk in obedience to his commandments.

The spiritual choices you and I make for our feet bring either light or darkness into our lives. We've been commanded to "walk as children of light" (Eph. 5:8) and to "walk in the light, as he is in the light" (1 John 1:7). We do this as we follow the Church leaders and obey the gospel. As the Savior promised, "He that followeth me shall not walk in darkness, but shall have the light of life" (John 8:12).

On the other hand, "The fool walketh in darkness" (Eccl. 2:14) and "leave[s] the paths of uprightness, to walk in the ways of darkness" (Prov. 2:13). How do the foolish and wicked use their feet to "walk in darkness"? By disobedience. "If you keep not my commandments, the love of the Father shall not continue with you, therefore you shall walk in darkness" (D&C 95:12). Think about that great principle. *Anything* we do with our bodies to separate ourselves from feeling and experiencing the love of the Father will cause us to walk in darkness instead of in the light of Christ. The damage this does to one's life is so serious that Christ called it a "very grievous sin" to "walk in darkness at noon-day." (See D&C 95:6.) Surely, the way we use our feet has serious consequences for our spiritual lives.

SLOW FEET AND FAST FEET

"Flee also youthful lusts" (2 Tim. 2:22).

Feet can move our bodies quickly or slowly. Proverbs speaks of six things which the Lord actually hates and they all have to do with the misuse of various parts of our bodies. One of them is "feet that be swift in running to mischief" (Prov. 6:18). Isaiah also spoke of those whose "feet *run* to evil ... and their thoughts are thoughts of iniquity" (Isa. 59:7). There's always a link between the thoughts of our mind and the actions of our feet—or any other parts of our bodies. Thus Paul challenged us to "walk not after the flesh, but after the spirit" (Rom. 8:1) and promised that those who do so "shall not fulfil the lust of the flesh" (Gal. 5:16).

Many times, the speed at which our feet move has a lot do with our spirituality. Slow feet, for example, might be considered synonymous with slothfulness or procrastination. When Mary told the apostles that Christ's body was missing and that the tomb was empty, they didn't take a leisurely stroll to investigate, but used their feet to *run* to the tomb with great haste. The question, in our analysis of spiritual anatomy is, How do we use *our* feet in the Lord's service? Are we swift to do our duty, or do we hold back and move reluctantly and slowly?

One day a man was arrested at a local high school for exposing himself to the students. It was an ugly sin, but he was deeply and sincerely remorseful. He acknowledged his sin to the Church. He committed himself to long-range therapy. He even went to the homes of the offended students and apologized. Almost everyone rallied around this brother with forgiveness and support, yet he could not find God's forgiveness because he could not forgive himself.

One Sunday, as the sacrament was being passed, this man left the meeting and almost ran from the building. There could be no doubt in the minds of the congregation as to how he felt, or why he left. As disciples of the Master, who said, "I am come to seek and to save that which is lost" (Luke 19:10), what should the feet in that congregation have done when this man left? Think how wonderful it would have been had the entire congregation leaped to their feet and gone after this hurting brother.

Fortunately, one man did. When he got outside the church, he saw that the man was already at the end of the block, literally running down the street. And, bless his heart, he chased after the man and cornered him in a cul-de-sac. He said, "Please come back. We want you with us."

The man was sobbing. "No, I'm not worthy to be there," he said. This spiritually sensitive man, who used his feet to swiftly pursue a brother in need, said, "Neither are *we!*"

While speedy feet are often important, sometimes an overemphasis on speed can be damaging as we try to do too much too soon. As we attempt to lengthen our stride, we mustn't confuse always going fast with being valiant. Often, what we need most isn't speed, but patient, steady growth. The strength of one's commitment to "endure to the end" is often more important than the pace of one's growth. As Elder Marvin J. Ashton counseled, "The speed with which we head along the straight and narrow path isn't as important as the direction in which we are traveling. That direction, if it is leading toward eternal goals, is the all-important factor" (*Ensign*, May 1989, p. 21).

Let's remember that God didn't create the world in one session, but divided his labors into progressive steps. We're expected to do the same in our process of personal development. In our eagerness to please the Lord, we mustn't expect too much of ourselves too soon and then fall into depression because we're not perfect as quickly as we'd hoped to be. We know the Lord understands that salvation and perfection are not achieved in a single *event*, but by a *process* of growth, requiring lots of practice.

Jesus said that we should "not run faster or labor more than you have strength and means" (D&C 10:4) and that "ye must *practice* virtue and holiness before me continually" (D&C 46:33). The word "practice" shows that God is allowing us time to learn and grow and that we should "continue in patience until [we] are perfected" (D&C 67:13). "Let us run with patience the race that is set before us" (Heb. 12:1).

While we're encouraged to ever be striving forward in the path to exaltation, there are also times when we're asked to stop our feet entirely and actually stand still. Must our feet and minds always be on the go, or do we have the spiritual ability to pause occasionally and reflect? "Stand still and consider the wondrous works of God" (Job 37:14). None of us would want to offend God, yet he has warned us that this can happen when we "confess not his hand in all things" (D&C 59:21). How sad it is when our feet go rushing through life so quickly that we don't take time to pause and notice the majesty of God as manifest in the details of his creations, or to notice his influence in our daily affairs!

Still another reason for standing still occasionally is that sometimes we're so busy rushing about, trying to do all that we can to meet our objectives, that we block out the divine assistance the Lord is trying to add to our efforts. Sometimes we can't recognize his answers to our prayers because we're so filled with fear, doubt, or frantic activity that we haven't trained our spiritual feet to "stand still" and discover what his power could do in our lives. For example, we can imagine the panic the Israelites must have felt as the 600 chariots from Egypt approached, trapping them against the Red Sea. It's easy to picture them running frantically about the camp searching for an escape, though there was none. "Fear ye not," was the word of Moses. "*Stand still, and see* the salvation of the Lord" (Ex. 14:13; see also 2 Chr. 20:17). "Therefore, dearly beloved brethren, let us cheerfully do all things that lie in our power; *and then may we stand still, with the utmost assurance,* to see the salvation of God, for his arm to be revealed" (D&C 123:17).

FEET THAT CAN WITHSTAND TEMPTATION

"Wherefore take unto you the whole armour of God, *that ye may be able to withstand in the evil day,* and having done all, to stand" (Eph. 6:13; see also D&C 27:15). What does spiritual armor have to do with our feet and the ability to withstand temptations? Ancient warriors always protected their feet along with the rest of their bodies. The scriptures about armor counsel us to do the same by having our feet "shod with the preparation of the gospel of peace" (Eph. 6:15). The scriptures contain more than 500 discussions on being prepared. Having our feet prepared to respond to various situations is a vital part of spiritual armor, and this means deciding *ahead of time* how we'll respond when Satan presents his temptations to misuse our bodies. Unless we prepare our response in advance, it's difficult to refuse temptations at the time of presentation, when we're caught in the strong pull of desire to give in and indulge. If we wait until that moment of passion to decide whether or not to give in, we often lose the battle, because our choices will be swayed by the desires we *feel* more than by what we *believe* and know we should do.

Proverbs 22:3 tells us that "a prudent man *forseeth* [anticipates] the evil, and hideth [prepares] himself: but the simple pass on, and are punished." "Passing on" means plunging into the tempting circumstances unprepared. Punishment comes naturally, by the consequences of unwise choices and refusal to prepare against the likelihood of temptation.

In war, a general counsels with his staff by asking, "What will we do if the enemy does *this*? What will we do if he does *that*?" One of the most important strategies for engaging in spiritual combat and making proper choices for our bodies is preparing ourselves to meet our temptations ahead of time, *before* the crises of desire arrives. As Jon M. Taylor said:

> A term commonly used by safe-driving experts is "defensive driving," implying the recognition that other cars on the road are a potential threat and that *hazardous situations should be anticipated.* Waiting to react until a danger is actually present may be too late.
>
> So it is with temptation. Once we decide we want to do right and then pray for the constant influence and prompting of the Holy Ghost, *we can anticipate situations that could sorely tempt us. We can then look ahead and be prepared.* It is during our stronger moments that we should prepare a plan of action for the moment of temptation *(The New Era*, Nov. 1972, p. 38).

Joseph resisted a series of enticing temptations in Egypt because he prepared himself ahead of time. Potiphar, an officer of Pharaoh and a captain of his guard, purchased Joseph as a slave and eventually placed him in charge of all his affairs. Potiphar's lonely wife was attracted to Joseph, and she "cast her eyes upon Joseph; and she said, Lie with me. But he refused, and said unto his master's wife . . . There is none greater in this house than I; neither hath he kept back any thing from me but thee, because thou art his wife: how then can I do this great wickedness, and sin against God?" (Gen. 39:7-9).

Seldom does Satan remove or abandon a temptation simply because we refuse it the first time. Over and over he presents the suggestions, finding new and enticing ways to urge us to misuse our bodies, and so it was with Potiphar's wife. "And it came to pass, as she spake to Joseph day by day, that he hearkened not unto her, to lie by her, *or to be with her*" (Gen. 39:10).

We notice Joseph's wisdom. Not only did he refuse the act of adultery, but he refused to spend time alone with her, for it's far easier to *prevent* a temptation than it is to *resist it* once the opportunity for indulgence has arisen.

You know the story. There came a day when "Joseph went into the house to do his business; and there was none of the men of the house there within" (Gen. 39:11). This was the private opportunity she was seeking. After all her previous attempts had failed, she actually took hold of Joseph and once again insisted on having sex with him. Surely Joseph was prepared for this situation because he had contemplated it in advance and planned what he would do. "And she caught him by his garment, saying, Lie with me: and he left his garment in her hand, *and fled, and got him out*" (Gen. 39:12).

When Joseph used his feet to run from Potiphar's wife, I don't believe that was a spur-of-the-moment decision. Do you? Of course it could have been, but I believe he'd probably sat in his quarters pondering the situation and wondering just how far she would go. I believe he predetermined in his mind exactly what he would do if it came to a showdown. And because he had *predetermined* so definitely that he would not do this thing, when he was trapped by this woman's wiles, his only choice was to run—and he did.

There are times in our lives when Satan catches us in situations that we don't expect or foresee. But if our choices for righteousness have been predetermined, then we, like Joseph, must use our feet to swiftly remove ourselves from those dangerous situations. I believe this is what it means to have our "feet shod with the preparation of the gospel"—to decide ahead of time.

Of course it isn't only our *preparations*, but the way we actually *use* our feet to walk through life each day that determines our power to *stand* against temptation and adversity when the times of testing comes. When Joshua sent soldiers to conquer the people of Ai, their victory seemed so certain that he only sent a small portion of his available men. He was astonished when they were defeated. The reason for the unexpected defeat was that someone in the army stole some idols, which were, of course, forbidden. This interest in idolatry offended the Lord and he withdrew his support of the army. His explanation to Joshua teaches an important principle that applies to us as well:

"There is an accursed thing in the midst of thee, O Israel: *thou canst not stand before thine enemies, until ye take away the accursed thing from among you*" (Josh. 7:13).

Trying to overcome one bad habit while clinging to another favorite sin (or "accursed thing") is like trying to stand erect with each foot in a different boat. We can't expect to grow closer to our Father in Heaven in one area of our lives while clinging to secret sins in other parts. If we would stand tall and firm against our satanic enemies, we must be willing to remove from our lives *every* accursed thing and give ourselves totally to the Lord. Then, and only then, can he bestow the grace that enables us to "withstand the evil day" and conquer the things that would place barriers between us and him. As President Spencer W. Kimball emphasized:

> In abandoning sin one cannot merely wish for better conditions. He must make them. He may need to come to hate the spotted garments and loathe the sin. He must be certain not only that he has abandoned the sin but that he has changed the situations surrounding the sin. He should avoid the places and conditions and circumstances where the sin occurred, for these could most readily breed it again.
>
> He must abandon the people with whom the sin was committed. He may not hate the persons involved but he must avoid them and everything associated with the sin.
>
> He must dispose of all letters, trinkets, and things which will remind him of the "old days" and the "old times." He must forget addresses, telephone numbers, people, places and situations from the sinful past, and build a new life. He must eliminate anything which would stir the old memories *(The Miracle of Forgiveness* [Salt Lake City: Bookcraft, 1969], pp. 171-172).

"By this ye may know if a man repenteth of his sins," the Lord declared. "Behold, he will confess them *and forsake them*" (D&C 58:43). In other words, without a total forsaking of the sin and all its "accursed" connections, there's no real repentance, no matter how sorry we feel. "Thou canst not stand before thine enemies, until ye take away the accursed thing from among you" (Josh. 7:13). A person, for example, who stops committing *physical* adultery while continuing in mental imaginings and longing for an abandoned partner has not truly repented, even though the physical act is being avoided. Until the "accursed thing" that person is clinging to is rooted from the emotions and thoughts and replaced with more holy thoughts and desires, this individual will be vulnerable to temptations and continuing defeat.

> Therefore the children of Israel could not stand before their enemies, but turned their backs before their enemies, because they were accursed: *neither will I be with you any more, except ye destroy the accursed from among you*" (Josh. 7:12).

> The true spirit of repentance, which all should exhibit, embraces a desire to stay away from sin. *One cannot simultaneously be repentant and flirt with transgression* (Spencer W. Kimball, *The Miracle of Forgiveness* [Salt Lake City: Bookcraft, 1969], p. 215).

With the multitude of evil places our feet can take us in today's world, the Lord has asked us to "Stand ye in holy places, and be not moved, until the day of the Lord come" (D&C 87:8; see also D&C 45:32). Here's an important fact about behavior. While it takes time to control and purify our thoughts, we all have the muscular control to prevent our feet from physically taking us where our unworthy thoughts may direct. Controlling our feet gives us time to work on the spiritual problems. "Put on the whole armour of God, *that ye may be able to stand* against the wiles of the devil" (Eph.6:11).

DIVINE AID FOR OUR FEET

One of the specific things we've been asked to do with our feet is to magnify our callings. "Thou shalt stand in the place of thy stewardship," the Lord counseled, because "he that is slothful shall not be counted worthy to stand, and he that learns not his duty and shows himself not approved shall not be counted worthy to stand" (D&C 107:100).

We're often called—both in the circumstances of life and in church callings—to walk and serve in ways that astonish and frighten us. We question the ability of our feet to walk in fulfillment of the overwhelming assignments and new challenges. Yet our faith in Christ can carry us over the roughness and unfamiliarity of any terrain the Lord has called us to walk through. If Peter's untrained feet could find the power to walk on water simply because Jesus invited him to join him, then we, too, can walk over or through any circumstance the Lord calls us to. (See Matt. 14.)

Our feet get tired after just a couple of hours walking the shopping malls. Imagine the pain in the feet of the pioneers who walked a thousand miles to settle the Salt Lake Valley! How could they have done it if the Lord had not made them equal to the task, as he did for the Israelites, who walked for forty years before obtaining the promised land? "Yea, forty years dist thou sustain them in the wilderness, so that they lacked nothing; their clothes waxed not old, *and their feet swelled not*" (Neh. 9:21). We of the latter days are promised that obedience to the Word of Wisdom will result in a quickening of our feet when the trials of our time require physical or mental endurance beyond normal strength. "And [they] shall run and not be weary, and shall walk and not faint" (D&C 89:20).

OUR FEET IN THE DAY OF JUDGMENT

"For his eyes are upon the ways of man, and he seeth all his goings" (Job 34:21).

The Lord has also urged us to prepare our feet for the day of judgment. "And I would," lamented Mormon, "that I could persuade all ye ends of the earth to repent and to *prepare to stand* before the judgment-seat of Christ" (Morm. 3:22). If we don't use this life to repent and prepare, *"we will stand with shame* and awful guilt before the bar of God" (Jacob 6:9) because "ye cannot hide your crimes from God; and except ye repent they will stand as a testimony against you at the last day" (Alma 39:8).

"How will any of you feel," Alma challenged, "if ye shall stand before the bar of God, having your garments stained with blood and all manner of filthiness?" (Alma 5:22). His disobedient son later explained exactly what it feels like to face that judgment unworthy and unprepared. 'Oh, thought I, that I could be banished and become extinct both soul and body, that I might not be brought to stand in the presence of my God, to be judged of my deeds" (Alma 36:15). "But this cannot be; *we must come forth and stand before him* in his glory, and in his power, and in his might, majesty, and dominion, and acknowledge to our everlasting shame that all his judgments are just" (Alma 12:15).

No person can escape this requirement to stand before God and give an account of his or her life. "For we shall *all* stand before the judgment seat of Christ" (Rom. 14:10), even "every soul who belongs to the whole human family of Adam" (Morm. 3:20). "Watch ye therefore, and pray always, that ye may be accounted worthy to escape all these things that shall come to pass, and to stand before the Son of man" (Luke 21:36).

If we learn to use our feet (and the other parts of our bodies) properly, we won't be afraid to stand before God on that day of judgment. The gospel ordinances and process make it possible "that ye may be sanctified by the reception of the Holy Ghost, *that ye may stand spotless before me* at the last day" (3 Ne. 27:20). We are all promised, through our obedience and faithful service, "*that ye may stand blameless* before God at the last day" (D&C 4:2).

> The way is prepared that whosoever will may walk therein and be saved (Alma 41:8).

Chapter Eight
THE NATURAL MAN DILEMMA

One main reason we came to this school on earth was to let our spirits learn to control and use a physical body to produce joy and increased capacities instead of succumbing to its captivity and restrictions. This mortal probation is, in large measure, an endurance test to see if we can dwell inside a mortal body of fallen flesh without becoming slaves to it. It's a contest, if you will, to see which will win: the flesh or the spirit, the will of the body or the will of the person.

Three major factors prevent us from using our bodies wisely and living as righteously as we want to:

1. The ever-present and powerful influence of the satanic forces that seek to ruin our mortal probation.

2. The limitations imposed by our fallen, natural-man state.

3. The unnecessary limitations we impose upon ourselves with improper guilt for being mortal and imperfect.

Now let's discuss each of these handicaps, along with the solutions our loving Heavenly Father has provided.

THE OPPOSITION OF SATANIC ENEMIES

"Satan is insanely jealous of your body. He would like to possess and destroy it" (Carol J. Wood, *The New Era*, June 1978, p. 9). Many of the weaknesses and sins that trouble our conscience come not from our wicked choices, but because of the unseen influence of satanic enemies who are skilled at deceiving and manipulating mortals. "For

the devil is an enemy unto God, and fighteth against him continually, and inviteth and enticeth [us] to sin, and to do that which is evil
continually" (Moro. 7:12). Satan's war against Christ and all of us who
chose his side in the premortal world continues here on earth in ever-
increasing intensity, for "it was given unto him to make war with the
saints, and to overcome them; and power was given him over all kindreds, and tongues, and nations" (Rev. 13:7).

> That war, so bitter, so intense, has gone on, and it has never ceased. It
> is the war between truth and error between agency and compulsion,
> between the followers of Christ and those who have denied Him (Gordon
> B. Hinckley, *Ensign,* November 1986, p. 42).

> This is not a war for territory or wealth; it is a contest for the eternal
> souls of men and women, boys and girls, the literal offspring of God, our
> Heavenly Father. But how many of us realize how serious this conflict is?
> Do we understand what the devil is trying to do to us? The scriptures warn
> that the devil will make war with the Saints of God. He will attack them
> with all the wicked devices his pornographic mind can devise (Mark E.
> Petersen, *Ensign,* May 1980, p. 69).

When war is fought between opposing nations, each army has two
objectives: kill or capture the enemy. When enemy forces are captured,
they're bound and placed in prison camps where they are controlled
and often tortured. Satan's goal is the same: to make us use our bodies in ways that will imprison us as slaves to bad habits and sins that
kill spiritually. He already failed to win us to his side as spirits in the
premortal world; now, in this physical world, his main strategy is to
capture us through the desires and vulnerabilities of our flesh. Richard
G. Scott pointed out that "Satan would convert divinely independent
spirits into creatures bound by habit, restricted by appetite, and
enslaved by transgression. He has never deviated from his intent to
enslave and destroy" (*Ensign*, Nov. 1981, p. 11).

Our modern society worships pleasure. It encourages and glorifies
pampering and personal indulgence. "If it feels good, do it. It's your body,
so you owe it to yourself to please yourself" is the urging of our modern
culture. Because of this treacherous philosophy, many well-meaning

Latter-day Saints have found themselves imprisoned within a body that demands continual indulgence beyond that which is spiritually and physically healthy. They find themselves compelled by the demanding and insistent desires of the flesh to repeat harmful habits over and over, in spite of what it does to family relationships, self-confidence, and spirituality. Instead of their bodies being a sacred temple where the Holy Ghost can dwell as a constant companion and where the Lord's Spirit is free to visit and nourish, they find themselves behind the bars of self-imposed captivity, subject to the constant companionship of evil spirits.

The scriptures contain many warnings of the dangers of misusing our bodies in ways that lead to satanic captivity. "Verily, verily, I say unto you, ye must watch and pray always, lest ye be tempted by the devil, and ye *be led away captive* by him" (3 Ne. 18:15). A *captive* is a prisoner, a person who has lost his freedom and is under the control of others. "The demons that take possession of men, overruling their agency and compelling them to obey Satanic bidding, are the unembodied angels of the devil, whose triumph it is to afflict mortals, and if possible to impel them to sin" (James E. Talmage, *Jesus the Christ* [Salt Lake City: Deseret Book, 1956], p. 183).

Whether we say yes to temptations *deliberately*, or only *unintentionally* through weakness, we're nevertheless in danger of being "taken captive by the devil, and led by his will down to destruction" (Alma 12:11). When we first give in to the lesser desires of our flesh, we are only creating a vulnerability or disposition to yield to future temptations. However, the more often we yield, the stronger Satan's influence becomes. "Why will ye yield yourselves unto him that he may have power over you?" (Alma 10:25).

Each time we give in to evil enticements and use our bodies in inappropriate ways, we open ourselves to the influence of Satan's demon tempters as surely as if we had posted a welcome sign. The demons we invite into our lives are highly skilled in helping us perpetuate those choices until we reach the point of enslavement and lose our agency. "It is as though Satan ties strings to the mind and body so that he can manipulate one like a puppet," said Elder Richard G. Scott (*Ensign*, May 1986, pp. 10- 11).

For example, those who allow their thoughts to wander into sexual fantasies will find their thoughts more and more difficult to control. Having opened the door to unworthy thoughts, they have unknowingly invited unclean devils to enter and take control of their minds. Those spirits will then guide those persons' thoughts into progressively downward spirals of filthiness, until the incessant thoughts compel them to seek wicked fulfillment by acting out the lewd desires. As Melvin J. Ballard said: "Secret weaknesses and vices leave an open door for the enemy of your souls to enter, and he may come in and take possession of you, *and you will be his slave*" (*The New Era*, March 1984, p.38). We're captured and imprisoned by satanic enticements as surely as any soldier in a prisoner of war camp. "Know ye not, that to whom ye yield yourselves servants to obey, his servants ye are to whom ye obey; whether of sin unto death, or of obedience unto righteousness?" (Rom. 6:16).

While Christ has the power and is anxious to lift us above our fallen nature (see Chapter Eleven), not everyone is willing to accept his precious gift. Many people prefer to remain in the carnal state rather than exercise the self-discipline required to rise above it. Thus, "he that persists in his own carnal nature, and goes on in the ways of sin and rebellion against God, remaineth in his fallen state and *the devil hath all power over him*" (Mosiah 16:5). "I say unto you, [how] can ye think of being saved when you have yielded yourselves to become subjects to the devil?" (Alma 5:20).

In his miserable existence, Satan's only reward, his only pleasure, is the victory he achieves when he persuades us to misuse our bodies. "And thus he goeth up and down, to and fro in the earth, seeking to destroy the souls of men," for "he seeketh that all men might be miserable like unto himself" (D&C 10:27; 2 Ne. 2:27). The prophets have pleaded with us to "remember the awfulness . . . of yielding to the enticings of that cunning one" and to remember that "to be carnally-minded is death, and to be spiritually-minded is life eternal" (2 Ne. 9:39).

"And because he [Satan] had fallen from heaven, and had become miserable forever, he sought also the misery of all mankind" (2 Ne. 2:18).

THE OPPOSITION OF FALLEN FLESH

The body with its five or more senses, with its appetites and passions, is essential to life and happiness, but in the ultimate analysis it is only a means to a higher end. When man makes its gratification an end in itself, he frustrates the purpose of life and sensuality (David O. McKay, *Pathways To Happiness* [Salt Lake City: Bookcraft, Inc., 1957], pp. 164-65).

Heavenly Father is the supreme intelligence in the universe. He is perfection, the ultimate of all that can be, both physically and spiritually. Our physical bodies, precious and sacred, are to be reverenced and respected, if for no other reason than they are created "in the image of his own body, male and female" (Moses 6:9). *But the bodies we dwell in now are not the same as the bodies he created.* When Adam and Eve fell and were cast from the Garden of Eden, we, their posterity, inherited lesser bodies—bodies that are naturally predisposed to be carnal, sensual, devilish, and in a state that is contrary to eternal happiness.

And now, my son, *all men that are in a state of nature,* or I would say, in a carnal state, are in the gall of bitterness and in the bonds of iniquity; they are without God in the world, and they have gone contrary to the nature of God; therefore, they *are in a state contrary to the nature of happiness* (Alma 41:11).

God did not create us as evil beings, nor did he place within us the desire to sin. This carnal nature came into our disposition because of the fall. It was Satan who "did beguile our first parents, *which was the cause* of their fall; *which was the cause* of all mankind becoming carnal, sensual, devilish, knowing evil from good, subjecting themselves to the devil" (Mosiah 16:3). As Alma explained, when Adam and Eve were sent out from the Garden of Eden "they became fallen man . . . cut off from the presence of the Lord" (Alma 42:6-7). "Therefore," he said, "they had become carnal, sensual, and devilish, *by nature*" (Alma 42:10; also Mosiah 3:16).

This doesn't mean that Adam and Eve were wicked, because they weren't. They were noble and valiant spirits, and we honor them for choosing the difficult path that made it possible for us to follow them into mortality. But it does mean that they and every spirit to follow them into physical, mortal bodies become *subject to the carnal nature* that curses mortal flesh with the desires and tendencies of the "natural man." As Paul said, when we are compared to Heavenly Father's perfection, "all have sinned, and come short of the glory of God" and "there is none righteous, no, not one" (Rom. 3:23,10). And the Savior declared, "*All* flesh has become corrupt before my face" (D&C 112:23) and "the imagination of man's heart *is evil from his youth*" (Gen. 8:21).

This means that with the handicap of our carnal, fallen natures, and in comparison to God's perfect nature, no matter how righteously we live, as long as we dwell in a body of mortal, fallen flesh, we're subject to and responsible for controlling and overcoming the carnal nature of our fallen flesh, for "as in Adam, *or by nature,* they fall" (Mosiah 3:16). President Spencer W. Kimball said:

> The way to eternal life is clear. It is well marked. It is difficult. Evil and good influences will be ever present. One must choose. Generally, *the evil way is the easier,* and since man is carnal, that way will triumph unless there be a conscious and a consistently vigorous effort to reject the evil and follow the good *(The Miracle of Forgiveness* [Salt Lake City: Bookcraft, 1969], p 15).

Why did President Kimball say it's easier to sin than not to sin? Because there's a constant war between *our* will—that is, the will of our spirit entities—and the will of our mortal, fallen bodies of flesh, which have desires of their own, entirely independent of our own personal will and choice. (See 2 Nephi 10:24.) Anyone who has seriously attempted to alter physical habits (such as overeating, bad temper, lust, and so forth) has encountered the strength of this will of the flesh. Thus Lehi challenged his sons to "look to the great Mediator, and hearken unto his great commandments . . . and not choose eternal death, *according to the will of the flesh* and the evil which is therein" (2 Ne. 2:28-29).

We live in bodies that are strange and unfamiliar, bodies that want to make us do things that hinder our progress. "Dearly beloved," Peter implored, "I beseech you as strangers and pilgrims, abstain from fleshly lusts, *which war against the soul*" (1 Pet. 2:11). James also understood the struggles we face with the unworthy desires of our fallen flesh. "From whence come [these] wars and fightings among you?" he asked. "Come they not . . . of your lusts that war in your members [meaning parts of our bodies]?" (James 4:1). It's far better to "hunger and thirst" after righteousness than it is to hunger and thirst to fulfill your lust. Out-of-control desires are terrible taskmasters. "For the flesh lusteth against the Spirit, and the Spirit against the flesh: and these are contrary the one to the other: *so that ye cannot do the things that ye would*" (Gal. 5:17).

Because of this conflict, Paul said, "I know that in *me* (that is, in my flesh), dwelleth no good thing: for to will [choose] is present with me; but how to perform that which is good I find not" (Rom. 7:18). Paul is describing the frustration we all feel when we choose the right, make resolutions and covenants, only to find ourselves failing to live those choices as honorably and exactly as we planned, because the ever-present pull of the natural man holds us back in spite of our best efforts. Nephi expressed this frustration eloquently: "O wretched man that I am! Yea, my heart sorroweth because of my flesh; my soul grieveth because of mine iniquities" (2 Ne. 4:17).

Whether we spend our mortality in bodies that are temples for our spirits or in addictive prisons depends largely on the way we respond to our desires. As the brother of Jared said: "Now behold, O Lord . . . we know that thou art holy . . . and that we are unworthy before thee [and that] because of the fall *our natures have become evil continually*" (Ether 3:2). Paul confirmed this evil nature with his converts in Ephesus when he reminded them that before their conversion and repentance, they, like most of the world, indulged in "the lusts of [their] flesh, fulfilling the desires of the flesh and of the mind; and were *by nature* the children of wrath, even as others" (Eph. 2:3).

This frustrating condition, called being a "natural man," is so contrary to God's nature and perfect environment that the scriptures actually describe us, in our present condition, as *enemies* to God. King Benjamin, for example, taught that "the natural man is an enemy to God, and has been from the fall of Adam, and will be forever and ever, unless he yields to the enticings of the Holy Spirit, and putteth off the natural man and becometh a saint through the atonement of Christ the Lord" (Mosiah 3:19; see also Mosiah 2:37-38). (We'll discuss this process in the last chapter.) It's easy to understand James's words that "whosoever . . . will be a friend of the world is the enemy of God" (James 4:4). But how can *we*, the members of the Church, the children of God who are trying to obey his commandments—how can *we* be his enemies when we have not chosen the way of the world, and are not deliberately seeking to overthrow his work?

It isn't that our fallen status diminishes God's love for us, because nothing can do that. It is not *we*, ourselves, or our imperfections that make us his enemies, but our temporarily fallen natures, which are at war with all that God has planned for us. We're enemies in the sense that the natural, carnal state that we're in because of the fall is contrary to godliness and holiness and all that's necessary to bring us back to God, for "the carnal mind is enmity against God: for it is not subject to the law of God, neither indeed can be" (Rom. 8:7). What's so terrible about mortal flesh? What's terrible is the evil that fallen flesh leads us into when not restrained by gospel values. As you read the indictment to follow, judge it by what you know of history and see if you don't agree these are indeed the natural consequences of unguided, unrestrained mortality:

> Now the works of the flesh are manifest, which are these; Adultery, fornication, uncleanness, lasciviousness,
> Idolatry, witchcraft, hatred, variance, emulations, wrath, strife, seditions, heresies,
> Envyings, murders, drunkenness, revellings, and such like: of the which I tell you before, as I have also told in time past, that they which do such things shall not inherit the kingdom of God (Gal. 5:19-21).

All these natural-man tendencies make the unchanged, not-yet-born-again person an enemy to God, "so then *they that are in the flesh cannot please God*" (Rom. 8:8) until they are transformed, as Paul hastened to assure the Romans in the next verse. "But ye are not in the flesh, [even though still mortal] but in the Spirit, if so be that the Spirit of God dwell in you" (Rom. 8:9). Anything that can hinder God's divine plans for his children is an enemy to God. We're enemies in our fallen state because, if we *remain* in this state, we'll never be with him again. That's why he sent Jesus to rescue us, as we'll discuss in the final chapter.

We prefer to think of evil and temptation as coming from outside ourselves so that we don't have to face the reality of the carnal and fallen nature that dwells within every mortal body of flesh. While it's true (as we learned in the previous section) that many temptations do come from Satan and from our culture and environment, it's also true that those suggestions wouldn't entice us unless the seeds of carnality were already there within our fallen nature. As James said, "Every man is tempted, when he is drawn away *of his own lust*, and enticed" (James 1:14). When we allow sensory stimulations to rattle around in our minds and hearts, giving them contemplation and weighing the temporary pleasures they might offer against the difficulties and consequences they could produce, we're giving them time to incubate. This incubation period is very dangerous, for "when lust hath conceived, it bringeth forth sin: and sin, when it is finished, bringeth forth death" (James 1:15).

We fought a war for the privilege to come here and prove that we (the premortal spirits) could take charge of the flesh and rise above its lustful desires, but the spiritual life doesn't come naturally to fallen man. Jesus said, "My commandments are spiritual; *they are not natural . . . .*" (D&C 29:35). In other words, the things Christ asks us to do in obedience to the gospel do not come naturally to our fallen nature. They require a commitment to rise above our nature, to reach out of our natural selves, to deny ourselves and our natural tendencies. That's why Paul said, "The natural man receiveth not the things of the Spirit of God: for they are foolishness unto him: neither can he know them, because they are spiritually discerned" (1 Cor. 2:14).

LEARNING TO SEPARATE

When we travel into the space above the earth's atmosphere (about seventy-five miles up), we must protect our bodies with an artificial environment. It took scientists several decades to develop spacesuits suitable for providing this protection. A spacesuit is a complete miniature world—a self-contained environment that must supply everything an astronaut needs for survival and comfort. For example, the spacesuit must not only supply oxygen, but also remove the carbon dioxide and water vapor created as the astronaut breathes, as well as disposing of (or temporarily storing) other bodily wastes.

The spacesuit must insulate the wearer from the temperature extremes of space. Without the earth's atmosphere to filter the sunlight, the side of the suit facing the sun could be heated to a temperature as high as 250° F., while the other side, exposed to the darkness of deep space, may get as cold as minus 250° F. The spacesuit must also provide a pressurized interior to keep body fluids in the liquid state; otherwise, an astronaut's blood would boil. (At sea level, the air pressure on our bodies is about 14.7 pounds per square inch [psi]. Spacesuits for the space-shuttle era are pressurized at 4.3 pounds psi, while those designed for the space-station era will be pressurized to 8.3 psi.)

The spacesuit must protect the astronaut from radiation, provide protection for the eyes in an environment in which there is no atmosphere to absorb the sun's rays, provide facilities for speech communication, and perform many other functions to keep the person inside the suit alive and able to perform the duties of his or her mission.

An astronaut would never confuse his *spacesuit* with his *personal identity*. He's aware at all times that the spacesuit he wears is merely a covering for the real person inside. *And so it is with our physical bodies.* Think of your physical body as something separate from you—like a spacesuit worn by your spirit in this foreign environment of mortality. The body of flesh is not *you*; it's a suit you wear. When you die, you'll temporarily discard that suit as surely as an astronaut sheds his spacesuit when his mission is completed. You and I fought a war for the privilege of coming here to dwell in physical bodies during this Earth school. The problem is, we all have defective spacesuits—bodies that have a will of their own, bodies constantly trying to pull us away from thoughts and experiences that promote spiritual growth.

We see this separation of identity between our spirit entities and our physical bodies in Ezekiel's vision of the dry bones, "which were to be *clothed upon with flesh*, to come forth again in the resurrection of the dead" (D&C 138:43). We also see it in the Biblical statement, "Thou hast *covered me* in my mother's womb" (Ps. 139:13). Speaking of the coming of Christ to dwell in a physical, mortal body, King Benjamin said that "the Lord Omnipotent . . . shall come down from heaven among the children of men, and shall *dwell in* a tabernacle of clay, and shall go forth amongst men" (Mosiah 3:5). This means that Christ, an unembodied spirit (like we used to be) would come here, as we do, and *he*, the person, would be "covered" in the womb ("clothed upon with flesh") as his physical body was formed, to temporarily live in or inhabit this spacesuit-tabernacle, or body of flesh. Christ later substantiated this fact when he said, "I was in the world *and made flesh my tabernacle*, and dwelt among the sons of men" (D&C 93:4).

Melvin J. Ballard made it very clear how important it is to recognize the difference between ourselves and the bodies we wear:

> The greatest conflict that any man or woman will ever have will be the battle that is had with self. I should like to speak of spirit and body as "me" and "it." "Me" is the individual who dwells in this body, who lived before I had such a body, and who will live when I step out of the body. "It" is the house I live in, the tabernacle of flesh and the great conflict is between "me" and "it" *(The New Era,* March 1984, p. 35).

When your car is low on gas, you don't identify with *it* and say, "*I* am thirsty." You say, "My car needs some more gas." When I say things like "I'm tired, I'm hungry, I have a sore thumb," I'm identifying myself with the spacesuit I live in. This identification can be a deadly mistake, because it sets up a subconscious obligation to satisfy the demands of the body instead of choosing the best course of spirituality.

We give ourselves a tremendous spiritual advantage in the battle against the flesh when we learn to separate ourselves from the houses of flesh we live in. For example, the struggle with weight gain or sexual desires seems quite different when we say to ourselves that "my body" is hungry or thirsty, or has certain desires, instead of saying "*I*" have these needs.

The space suit is a servant to the astronaut, not his master. In the same way, our bodies should be our servants, not our masters. "And why should *I* [that is, *me*, the spirit entity inside the body] yield to sin, because of my flesh?" Nephi asked. "Yea, why should *I* give way to temptations, that the devil have place in my heart to destroy my peace and afflict my soul?" (2 Ne. 4:27). We see this scriptural clarification of the conflict between our chosen path and the fallen, carnal desires of the flesh in Paul's words: "Let not sin therefore reign in your mortal body, that *ye* should obey *it* in the lusts thereof" (Rom. 6:12). Paul didn't say, "Don't give in to *your* lustful desires." He was urging us to recognize the separation.

How long would you wear a pair of shoes that forced you to go where *they* chose to go rather than where *you* wanted to go? What would you do with a car that refused to follow your directions—a car that went faster, for example, when you stepped on the brake, or turned left every time you tried to turn right? How would you like to live in a home that made up its own mind when the windows or doors would be open or shut, when the lights should be on or off, whether the temperature should be cooled or warmed? Silly, isn't it? Yet how many of us allow the natural desires and passions of our physical body to dictate where we go, what we feel, what we eat or drink, what we watch or do—to our everlasting damnation?

If we are to make temples of our bodies instead of prisons, it's crucial to realize that *we*, the spirits within the bodies, are facing a war between *ourselves* and our mortal housing. This war is confusing, and we'll have many occasions to blame ourselves for natural, carnal desires. But if we place all the blame on *ourselves*, assuming our weakness is because we are filthy, wicked, or uncommitted, we will defeat ourselves. To win these battles between the will of the flesh and our own chosen priorities, we must learn to separate the two and realize that it isn't *me* fighting *myself*, but me fighting my fallen flesh.

CONQUERING THE FLESH

Jacob talked about people who choose the pleasures of the flesh over spiritual progression, people who "grow hard in their hearts, and *indulge themselves* somewhat in wicked practices" (Jacob 1:15). Rather than face all the complications of secrecy and shame that accompany such choices, the Lord has asked us to "entangle not yourselves in sin, but let your hands be clean, until the Lord comes" (D&C 88:86).

We live in a world so full of wickedness that it's probably impossible to avoid exposure to lustful stimulations and suggestions. But it's quite possible to stop those enticements from taking root in our emotions and giving birth to inappropriate actions. This is the major challenge of mortality, for "we have a labor to perform whilst in this tabernacle of clay, that we may conquer the enemy of all righteousness, and rest our souls in the kingdom of God" (Moro. 9:6).

In the premortal life, we lived a long time as spirits without bodies of flesh. Now that we have to struggle so hard to conquer the flesh, we sometimes wish we could be spirits again. But what we need is not *escape* from the flesh, but *victory* over it, for those who die and temporarily set aside their physical bodies lose so many forms of feeling and expression that they look "upon the long absence of [their] spirits from [their] bodies to be a bondage" (D&C 45:17; see also 138:50). Paul spoke frequently on this subject. "I keep under my body," he said, "and *bring it into subjection*; lest by any means, when I have preached to others, I myself should be a castaway" (1 Cor. 9:27). Melvin J. Ballard said:

> I am charged to take possession of this house, this mortal tabernacle, and it is to be my servant. I am not to abuse it but keep it vigorous, clean, healthy, and strong. This exercise of controlling it once a month, that it must fast, is a healthy exercise of spiritual control over the material.
>
> If I can do this with regard to food, [then,] when this body craves something that is positively hurtful, then I have obtained power to say: "You cannot have it." Thus spiritual control over the body in all its activities, may be secured, beginning with control over the appetite (*Sermons and Missionary Services of Melvin J. Ballard*, n.d., p. 157).

Our mission to gain mastery over our bodies of flesh is so vital that the scriptures frequently link it to the Savior's crucifixion. For example, Jesus said, "If any man [or woman] will come after me, *let him deny himself, and take up his cross*, and follow me" (Mark 8:34). He also warned, "*He that taketh not his cross*, and followeth after me, is not worthy of me" (Matt. 10:38) and that "*whosoever doth not bear his cross*, and come after me, cannot be my disciple" (Luke 14:27). Inasmuch as acceptable discipleship hinges on the cross—both what he did on *his* and what he now expects us to do with *ours*—it behooves us to be certain we understand this concept.

When the scriptures tell us to take up and bear our personal crosses, they refer to the life-long duty to master the inappropriate, carnal desires of our flesh so that the spirit entity inside is the one in charge. For example, Paul said, "They that are Christ's have *crucified the flesh* with the affections and lusts" (Gal. 5:24). This self-denial is a necessary precursor to receiving Christ's transforming power to change our carnal nature to the spiritual nature that enables us to enjoy the fruits of the Spirit. We see the same linkage between mastery over the flesh and the symbolic bearing of our personal crosses in Alma's words to his adulterous son:

> Now my son, I would that ye should repent and forsake your sins, and go no more after the lusts of your eyes, *but cross yourself in all these things*; for except ye do this ye can in nowise inherit the kingdom of God. Oh, remember, and take it upon you, and *cross yourself in these things* (Alma 39:9).

It would be wonderful if we could conquer the desires of the flesh with one commitment, but this challenging struggle requires constant vigilance and enduring to the end. We know Christ recognized this when he said, "If any man will come after me, let him deny himself, and take up his cross *daily*, and follow me" (Luke 9:23). He wouldn't have acknowledged this challenge as being "daily" if he thought we could make one sincere vow and have done with it.

We all grow weary with the daily crosses we must bear. We can gain determination and commitment to bear those daily sacrifices by following the example of Christ. One thing that gave him the will and commitment to bear the awful agony of the atonement was the joy that he could foresee on the other side of the terrible suffering. We, too, can bear our burdens when we know it's temporary and well worth the rewards to come. Thus Paul offered this encouragement:

> "Let us lay aside every weight, and the sin which doth so easily beset us, and let us run with patience the race that is set before us, Looking unto Jesus the author and finisher of our faith; who for the joy that was set before him endured the cross" (Heb. 12:1-2).

Keeping the perfect joy and eternal rewards we can win by conquering unworthy desires of the flesh will give us the strength to deny immediate pleasures for the greater, everlasting pleasures to be had by righteous obedience.

MAKE NOT PROVISION

"If I regard [make place for] iniquity in my heart, the Lord will not hear me" (Ps. 66:18).

We all face the challenge of powerful desires. But if we do not learn to control them, we can make our foreordained struggles here far worse than they need to be. Just as Adam "*became subject to the will of the devil,* because he yielded unto temptation" (D&C 29:40), we too can give Satan additional power over us (to stimulate and amplify our natural desires) by making deliberate choices to yield to temptation and gratify inappropriate desires. As President David O. McKay said, "Let your needs rule you, pamper them [and] you will see them multiply like insects in the sun. The more you give them, the more they demand" (*Pathways To Happiness,* [Salt Lake City: Bookcraft, Inc., 1957], p. 110). "Be faithful, and yield to no temptation" (D&C 9:13).

When Alma taught his son to "see that ye bridle all your passions, that ye may be filled with love" (Alma 38:12), he was affirming the need all people have to train and guide their emotions and desires so that they are kept within the bounds the Lord has given us. A bridle is a harness used to *restrain* and *guide* an animal, such as a powerful horse. If we do not guide and control our desires and passions, they will lead us into the captivity of selfishness. And selfishness destroys our ability to love, to feel concern and to give service to others because of preoccupation with our own pleasures.

Many people try to excuse sinful behavior by stating that they are just "doing what comes naturally." That's a quicksand philosophy because inappropriate pleasures are always fleeting, temporary and empty of lasting joy. Esau was not the only one to "sell his birthright" for a mess of pottage. (See Gen. 25:29-34.) "Do not spend money for that which is of no worth," Jacob counseled, "nor your labor for that which cannot satisfy" (2 Ne. 9:51). Finding ourselves possessed by a strong desire which we didn't consciously choose to feel doesn't justify acting on it. "Do not endeavor to excuse yourself in the least point because of your sins" (Alma 42:30).

Both Paul and Moroni warned of the dangers of indulging our flesh. "Be wise in the days of your probation," Moroni counseled as he explained that one of the wise uses of our bodies would be to "strip yourselves of all uncleanness" and to "ask not" or not seek for ways "that ye may consume it on your lusts, but ask with a firmness unshaken, that ye will yield to no temptation, but that ye will serve the true and living God" (Morm.9:28). Of course no one would actually utter a prayer for opportunity to fulfill their lust. "Ask not" in this verse means not to seek opportunities or ways to fulfill our lust, or as Paul expressed it, "Put ye on the Lord Jesus Christ, and *make not provision for the flesh*, to fulfil the lusts thereof" (Rom. 13:14). "Making provision" means arranging our time and affairs so that we may indulge unworthy desires outside the bounds the Lord has granted us.

Perhaps the best example I could give of "making provision" for the flesh was not a wicked one at all. It was something my mother had to do when I was a young boy and we lived in a little desert town with no cooler or air conditioner. Mom's health was poor, and the summer heat was very hard on her. I remember her creative "provisioning" to survive one week that was particularly hot and difficult for her. She froze some 7-Up until it was slushy; then, during the hottest part of the day, she opened the door to the refrigerator, sat in front of it with her bare feet inside, and sipped the almost frozen 7-Up. Now that is "making provision," and it got her through the ordeal.

But most of our "making provision" leads to trouble. For example, if we're trying to improve our physical health and lose weight, "making provision" for our physical appetites would be keeping the cupboards stocked with pastries, ice creams, candies and cookies, potato chips, soda pops and other types of unhealthy and fattening foods that modern society teaches us to crave. Not "making provision" would mean keeping fresh fruits and vegetables on hand to satisfy our hunger pangs and learning better ways of cooking. It's a matter of paying the price to reeducate our appetites. If we're trying to lose weight, it doesn't make much sense to stare at pictures of chocolate cakes for several hours a day. How much wiser to decide what we're going to eat before we ever enter the kitchen than to peer longingly into the refrigerator or cupboards looking for unhealthy foods.

If it's sexual lust we are trying to conquer, "making provision" might include searching the TV program guide for movies to stimulate our appetites and feed our cravings for sexual enactments. If our consciences won't allow us to actually look it up, another form of "making provision" would be to casually surf the cable or satellite channels to see what we might "stumble" onto. Of course, we do not usually admit to ourselves that we're actually searching for something to pollute our minds and increase our desires. Rather, we tell ourselves we're just checking to "see what's on." If we're truly committed to overcoming lustful desires, isn't this kind of "making provision" something like throwing gasoline on a fire you're trying to extinguish? How much wiser to first scan a program guide to find the wholesome, acceptable programs, avoiding the risk of being enticed by what we're likely to find on today's televesion channels! Say with the psalmist, "I have remembered thy name, O LORD, in the night, and have kept thy law" (Ps.119:55).

"Making provision" in dating would be arranging for private opportunities where something *could* happen if the partner were willing. A single person trying to control a lust for improper physical contact with a date must acknowledge those desires in advance and carefully plan dates to eliminate opportunities for improper intimacies: for example, never entering a bedroom together or engaging in prolonged parking in a dark car.

If masturbation is the lust of the flesh we are trying to overcome, "making provision" would be things like touching ourselves in a stimulating way, seeking visual inputs and fantasies that inflame our desires and make it difficult to resist, and seeking opportunities to be alone so that we can "do it" to ourselves. To change behavior *output* we must first change the *input*, so we stop feeding our mind with nudity and sexual portrayals. Not "making provision" for masturbation would be learning to keep our hands away from the private sexual parts of our body and controlling visual stimulations.

Satan's strategy for using our bodies to defeat us here in mortality is based on the immediacy of physical pleasure. Let's face it, if sin did not feel good, we wouldn't keep doing it. One of life's great challenges is learning to separate the harmful desires from the good ones. Before we decide to fulfill a desire, we must test it, we must measure it against gospel principles and determine if its fulfillment will be in accordance with our covenants and will benefit our progress.

Not "making provision," then, is facing the reality of the inappropriate desires of our flesh and planning strategies that will reduce or eliminate the *temptations* and *opportunities* to succumb. It's far better to eliminate the temptations *before* they occur than to struggle with them *after* our desires are aroused.

A member of our stake presidency told me how he learned to prevent himself from "making provision." His business requires him to travel, so he has to spend a lot of time in lonely motel rooms where he would certainly have the opportunity to sin in secret without being caught. Of course we never "get away with it" when we sin. If this great man ever took advantage of those opportunities, his spirituality would diminish, his relationship with his wife would be diminished, and his ability to fulfil his assignments would deteriorate as he lost the companionship of the Spirit. But the situation is there and he has a strategy. Paul warned, "Use not liberty for an occasion to the flesh, but by love serve one another" (Gal. 5:13). This brother didn't want to be unfaithful to his wife, even with mental adultery. So rather than having to deal with the temptations found on the TV, he just never turns it on, not even for the news. To fill the empty hours he takes books to read, studies the scriptures, and prepares talks and lesson assignments. He is serving both his family and the members of the stake by carefully guarding his virtue. He has grown past the temptations of "making provision" for the flesh by eliminating the possibility of even being enticed to do so.

May we each live our lives in such a way that it will never be said of us, as it was of the wicked King Noah, that "he did not keep the commandments of God, but he did walk after the desires of his own heart" (Mosiah 11:2). "Neither yield ye your members as instruments of unrighteousness unto sin: but yield yourselves unto God, as those that are alive from the dead, and your members as instruments of righteousness unto God" (Rom. 6:13).

When I think about controlling my flesh and not making provision to grant its lesser desires, I'm reminded of Lamoni's father, a cruel and wicked king. When touched by the Lord's Spirit, this man made a totally yielding and trusting commitment (which, by the way, is a necessity for all of us): "I will *give away all my sins* to know thee" (Alma 22:18).

One of my wife's pioneer ancestors illustrates the kind of commitment we must all achieve over the faults of our flesh. Her name was Mary Ann Richardson. She got an infection in her big toe that wouldn't go away, and she finally had to consult a doctor when it became gangrenous. He told her the toe would have to be amputated to save the foot and that he would cut it off for $100—a fortune in those days. She didn't have the money and told him she couldn't afford the operation. As she left, the uncaring doctor assured her that it would not be long before she would *have* to find the money and come back to him. However, he underestimated her resourceful determination. When she got home she took off her shoe and stocking, placed her toe on the chopping block, and removed the toe herself with a chisel and hammer. Are we willing to "give away" or to amputate from our behavior patterns the sins and bad habits that are rotting and festering our spirituality?

DON'T LET MISGUIDED GUILT DEFEAT YOU

All people, including the prophets and apostles, have desires and passions they must overcome. It's part of being human. Learning to live in physical flesh is one reason we came here. For example, when the superstitious people of Lystra were overawed by Paul's healing of a cripple and attempted to worship him as a god, he protested, saying, "Sirs, why do ye these things? *We also are men of like passions with you*" (Acts 14:15). Likewise, the great prophet Elias "was a man subject to like passions as we are" (James 5:17). Even Joseph Smith reported that after his first vision, "I frequently fell into many foolish errors, and displayed the weakness of youth, and the foibles of human nature; which, I am sorry to say, led me into divers temptations, offensive in the sight of God" (JS-H, 1:28).

If your car develops a mechanical defect, you don't condemn *yourself* for it. The car isn't *you*; it's just something you ride in. You get it repaired and get on with life. Unfortunately, however, Satan has become very skillful in defeating mortals by tricking us into identifying with the defects in our mortal tabernacles. By making us feel guilty for being human and imperfect, he diminishes our feelings of self-respect, self-worth, and confidence in our ability to rise above this fallen condition. We must not beat ourselves up because we have inappropriate desires. Nephi stated, "Not that I would excuse myself because of other men, but because of the weakness which is in me, according to the [fallen] flesh, I would excuse myself" (1 Ne. 19:6).

If any person ever wanted to make up for his mistakes and be all he could be for the Savior, it was Paul. Yet he described what we've all experienced from our fallen flesh when he said, "I find then a law, that, when I would do good, evil is present with me. For I delight in the law of God after the inner man [the spirit]: But I see another law in my members [the fallen flesh], warring against the law of my mind, and bringing me into captivity to the law of sin [the fallen nature] which is in my members [mortal body.] O wretched man that I am! Who shall deliver me from the body of this death?" (Rom. 7:21-24). In other words, no matter how sincerely I choose and plan to do right, no matter how much I hate to sin and disobey, I am weak and imperfect. I find it difficult to do all the righteous things I yearn to do for my Savior, and easy to slip and do the things I don't mean to do. I am constantly pulled down by the weaknesses and imperfections of my mortality, so that "what I would [do], that I do not; but what I hate, that do I" (Rom. 7:15).

But because he recognized the war between his fallen nature and his higher self, Paul never allowed his mortality to make him feel guilty or ashamed. Because he was committed to the Savior, he recognized, "It is no more *I* that do it, but sin [or the fallen nature] that *dwelleth in* me" (Rom. 7:17). All righteous Christians yearn to live a better spiritual life than their fallen nature allows them to. We constantly strive for a higher walk, but are continually pulled downward by mortal weaknesses of the flesh. Stephen E. Robinson has provided a helpful analogy:

> The effects of the Fall (physical, spiritual, mental, emotional, moral) constitute external handicaps that we must all work under in mortality, but we are promised that they will be removed from us in eternity through the victory of Christ. The compulsive and the perfectionists among us need to realize that a large part of why things go wrong in this life is the Fall—not their own incompetence.
>
> What is impossible is being totally separated from our carnal self. We are stuck, like missionaries with companions [our fallen body] who won't get out of bed. We can't just leave them. So we do the best we can to keep the rules ourselves while trying to make the companionship as productive as possible under the circumstances. And we look forward to a transfer. So in the flesh we do the best we can to resist the influence of our carnal companion and to keep the rules while trying to make our lives as productive as possible under the circumstances. And we look forward to the resurrection *(Following Christ* [Salt Lake City: Deseret Book Co., 1995], p. 63).

Because we all have to deal with inappropriate desires, the Lord has said that we will be judged not only by our desires, but also by our *response* to them. "And by their desires *and* their works you shall know them" (D&C 18:38). This means the judgment comes not because we *experience* unworthy desires, but on how we resist and rise above them. How comforting to know that we will be judged not by each mistake or imperfection alone, but also by the righteous desires we yearn toward and strive to live! Jesus said, "I, the Lord, will judge all men according to their work, according to the desire of their hearts" (D&C 137:9). I take this to mean that even if we are imperfect in our conduct because of human frailties, but are sincerely trying to do better, God will show mercy as he weighs our righteous desires against our mistakes and weaknesses. He is rich in mercy and patience towards our accountability because "he knoweth our frame [mortal body]; he remembereth that we are dust" (Ps. 103:14). As Elder Dallin H. Oaks said:

> When we have done all that we can, our desires will carry us the rest of the way. If our desires are right, we can be forgiven for the unintended errors or mistakes we will inevitably make as we try to carry those desires into effect. What a comfort for our feelings of inadequacy! *(Pure In Heart* [Salt Lake City: Bookcraft, 1988], p. 79).

Sometimes, when a particular part of our body urges us to satisfy lustful desires that are contrary to our covenants, we may wish we didn't have that particular member. But Christ said that "the body hath need of every member, that all may be edified together, that the system may be kept perfect" (D&C 84:10). Many of us don't like our bodies. They're too tall or too short, too fat or too thin. Some bodies are crippled, deformed, and ugly. Many of them don't function properly. They can be handicapped by disease or injury, or worn by age. I doubt if one single person has ever had a mortal probation exactly the way that person would have wished. Every person has some unwanted defect of character or of physical body. The complaints are endless.

You may feel trapped in your spacesuit body and wish you'd been issued a different one, but God doesn't make mistakes. Whatever problems we find in our "natural man" spacesuits, we have to learn to turn them over to the Lord and get on with life. As we pass through this life we either play the hand we're dealt or waste our mortality moaning and wishing things were different. Of course, we don't enjoy the problems we encounter here, but Lehi promised that the Lord "shall consecrate thine afflictions for thy gain" (2 Ne. 2:2). And Christ himself promised that "all things wherewith you have been afflicted shall work together for your good" (D&C 98:3). It's amazing what quantum leaps of progress people can make when they stop resenting their problems and start telling themselves, "What I have here in this defective spacesuit is a tremendous opportunity to grow and triumph."

When we were involved in school, we often grew weary of endless homework assignments. We wished we didn't have so much study and work to do. Brigham Young said, "Every trial and experience you have passed through is necessary for your salvation" (*1976 Priesthood Manual*, p. 228). Resenting the lessons necessary for our salvation is like paying a huge tuition to go to college and then resenting the homework. If we wish to rise above the natural man and reach for the Savior's image, we need to stop resenting our "homework" and start trusting the Lord, who is carefully choosing the lessons we need the most.

The two most powerful manifestations of the natural man are found in the carnal mind and the inappropriate desires of the heart. We will consider these next, and then conclude with the spiritual transformations made possible by the second birth.

Chapter Nine

THE MIND

For most of my life I considered the words "mind" and "brain" to mean the same thing, but I was wrong. The *mind* is the self-aware part of our eternal spirit that thinks and has consciousness, while the *brain* is the physical part of our body that carries out the instructions of our conscious, spirit mind. As Bruce R. McConkie said, "The sentient, conscious, and intelligent part of man—the part that perceives, feels, wills, and thinks—is called the *mind*" and "it is clear that the mind of man rests in the eternal spirit. Man's intelligence is in his spirit and not in the natural or mortal body" (*Mormon Doctrine* [Salt Lake City: Bookcraft, 1966], p. 501).

Heavenly Father designed and created the brain to be capable of operating our complex physical body on its own, automatically, with no conscious thought on our part. Do you wish to run, or jump, or walk into another room? You simply choose the action you want (with your spirit mind) and your physical brain takes over, controlling hundreds of muscles to propel your body and maintain proper balance, leaving your mind free to think about other things.

If you choose to run, for example, *you* don't need to consciously instruct your brain to accelerate the heartbeat or breathing rates. It will do that automatically for you. Your body is filled with tens of thousands of sensors that monitor and help the brain automatically adjust the chemical level of vital substances such as sugar, oxygen, and carbon dioxide in the blood, as well as physical conditions such as blood pressure and body temperature. When you eat food, for example, you don't need to think about or issue instructions to your digestive system. Your brain has been programmed to respond to the multitude of different foods you may place in your stomach and initiate any of millions of chemical responses to transform those substances into the energy you need.

The brain has specialized areas for dealing with multitudes of issues such as balance, hearing, vision, language, muscle control, taste, control of hormones, immune defenses, and so forth. Because the brain takes care of these millions of chemical and motion functions automatically for us, *we*—the spirit beings inside the body—are left free to think about the important things, like issues of morality, interaction with other people, and choosing those activities and feelings that move our spiritual progress forward. Imagine how little we could learn if our time had to be preoccupied each day with the maintenance and operation of our bodies!

We don't *think* with the physical brain alone. As mortals, we have two "control centers" that determine our eternal destiny as well as our present happiness or sorrows: the *mind* and the *heart*. These two centers work together inseparably, and it's impossible to dissect or consider the operation of one without the other. But in the scriptural sense, it's the *mind* that controls the thoughts and the *heart* which controls the emotions and desires. Whenever we sin or make mistakes, it's either because of wrong *thinking* or wrong *feeling*, or both because both the mind and the heart have been corrupted by the fall.

THE FALLEN MIND

Have you ever had a computer you couldn't trust because you could never count on it to do what you wanted it to do? When a computer stops functioning properly, it will either be because of defective hardware, or, most often, because the programing has become corrupted by a software glitch or virus. This is what's wrong with our mortal minds. We have, as the scriptures describe it, become "men of corrupt minds" (2 Tim. 3:8; see also D&C 33:4). This is because when we inherited the fall and the natural-man dilemma as described in the last chapter, it was not just our physical *bodies* that placed us in enemy status against God and godliness, but also our *minds* and *hearts*, or in other words, our thoughts and feelings which "began from that time forth to be carnal, sensual, and devilish" (Moses 5:13). "And God saw that the wickedness of man was great in the earth, and that every imagination of the thoughts of his heart was only evil continually" (Gen. 6:5).

It's not only our *outward* sins of behavior that trouble the Lord, but also our inward sins of *thought*, for "the thoughts of the wicked are an abomination to the Lord" (Prov. 15:26). For example, when God looked upon the wickedness of the people in Noah's time, it wasn't just their outward actions that disturbed him. He saw not only that "the wickedness of man had become great in the earth," but also that "every man was lifted up in the imagination of the thoughts of his heart, being only evil continually" (Moses 8:22). We can see the importance of clean thoughts to the Lord when the Bible tells us there are six things the Lord hates, and that one of them is "an heart that deviseth wicked imaginations" (Prov. 6:16-17).

Whether we experience mortality in bodies that are temples, or in bodies that become degrading prisons, depends not only on how we use the physical body, but also on what we allow to reside in our thoughts and imagination. All chosen sins begin in thought. The mind is the control center that's responsible for our mental choices. No thought is too small to have an effect because we always reap (that is, *become* or *do* or *fulfil)* in our *physical* lives what we've been sowing in our *mental* lives. If we think about an idea long enough, we're probably going to act it out, because the thoughts we focus on are like the blueprints of our future reality. Thus we are counseled that, if we would increase our spirituality, not only should "the wicked forsake his way," but "the unrighteous man his thoughts" (Isa. 55:7).

> Sow a thought, reap an act;
> Sow an act, reap a habit;
> Sow a habit, reap a character;
> Sow a character, reap an eternal destiny. (Author unknown)

Elder Joseph B. Wirthlin warned us not to trust the carnal mind when he said, "Experience teaches that when the will and the imagination are in conflict, the imagination usually wins" (*Ensign,* May

1982, p. 24). Because of our fallen traits, "the carnal mind is enmity against God: for it is not subject to the law of God, neither indeed can be" (Rom. 8:7.) That is why we "were sometime alienated and enemies in [our] mind by wicked works" (Col. 1:21).

Even though we were created in the image of our Heavenly Father, our fallen, carnal minds don't work the same way God's perfected mind does. "For my thoughts are not your thoughts, neither are your ways my ways, saith the Lord. For *as the heavens are higher than the earth*, so are my ways higher than your ways, and my thoughts [higher] than your thoughts" (Isa. 55:9). What a contrast! How high are the heavens above us? The nearest star is millions of light years away, and this is the Lord's contrast between his mind and ours!

When mankind fell and became subject to the will of the flesh and the will of the devil, one of the effects was that we became more selfish than unselfish. We became more oriented toward physical pleasures than spiritual values—more naturally self-willed than submissive to divine will. That's why "there is a way which seemeth right unto a man, but the end thereof are the ways of [spiritual] death" (Prov. 14:12). Thus, the Lord said, "I have spread out my hands all the day unto a rebellious people, which walketh in a way that was not good [because it was] after their own thoughts" (Isa. 65:2). Because of this fallen mental condition of our minds, we "are not able to abide the presence of God now . . . neither can any natural man abide the presence of God, neither after the carnal mind" (D&C 67:13, 12).

RISING ABOVE THE FALLEN MIND

As disciples, we strive to rise above such thought patterns while we learn to love and care about others at least as much as ourselves. We give serious heed to King Benjamin's warning: "I cannot tell you all the things whereby ye may commit sin; for there are diverse ways and means, even so many that I cannot number them. But this much I can tell you, that *if you do not watch yourselves and your thoughts . . . ye must perish*" (Mosiah 4:29-30). In this chapter, we'll focus on controlling and elevating our thoughts; in the next chapter, we'll focus on controlling and elevating the improper desires of the heart. As with all other parts of our bodies, there are specific things the Lord has asked us *to do* with our minds, and specific things he has asked us *not* to do.

PAY ATTENTION

Mortality is not just an arena in which to choose between good and evil, but also a time in which to learn spiritual priorities and teach our minds to focus our time and attention on those things that matter the most. One great challenge in perfecting our spiritual anatomy is the constant distractions and diversions competing for our attention. For example, have you ever tried to teach a class of people not really paying attention to you? They sort of listen, but your efforts aren't very fulfilling because you know their real attention is somewhere else. Perhaps that's a bit like what Heavenly Father experiences when he puts us here in this earth school to learn the important lessons we need to learn, but he can't seem to get our full attention.

Because those things we hold in the focus of our attention have such a powerful influence on our feelings and actions, the Lord has given hundreds of revelations on how we should use our thoughts and memories. He has asked us, for example, to take this preparatory life seriously, not to waste our lives or our precious minds in frivolity, but to "think soberly" and to "cast away [our] idle thoughts . . . far from [us]" (Rom. 12:3; D&C 88:69). He has also asked us to "cease from all your light speeches, from all your laughter, from all your lustful desires, from all your pride and light-mindedness" (D&C 88:121) and to "let the solemnities of eternity rest upon your minds" (D&C 43:34). He asks these things in the hope that we'll learn to direct our attention to the important issues of life, rather than drifting through mortality preoccupied with things that don't matter and that won't prepare us for eternity. Such instructions help us apply to ourselves the chastisement David Whitmer received when the Lord said to him, "Your mind has been on the things of the earth more than on the things of me, your Maker, and the ministry whereunto you have been called" (D&C 30:2). Mistakenly assuming that Pahoran was neglecting the desperate needs of his troops, Moroni sent him a scathing complaint which could easily be asked of *us* by the Lord concerning the duties of our mortal probation:

> And now behold, we desire to know the cause of this exceedingly great neglect; yea, we desire to know the cause of your thoughtless state. Can you think to sit upon your thrones in a state of thoughtless stupor, while your enemies are spreading the work of death around you? (Alma 60:6-7)

Of course, students pay a lot more attention to their teachers when they know they have just a short time to learn the material before an important test. You and I would probably pay a lot more attention if we'd remember that we have but a short time in this earth-school to learn our lessons and be "proven in all things" (See Abr. 3:25) before we face final exams in the day of judgment.

STOP FORGETTING

"Only *take heed* to thyself, and keep thy soul diligently, *lest thou forget* the things which thine eyes have seen, and *lest they depart from thy heart* all the days of thy life" (Deut. 4:9). This scripture contains the focus of God's concern for fallen man's memory, because we are so "swift to do iniquity but *slow to remember* the Lord [our] God" (1 Ne. 17:45). Unfortunately, this misdirected preoccupation and forgetfulness is part of being a natural man and something we must all struggle to overcome. As Mormon lamented:

> We can behold how false, and also the unsteadiness of the hearts of the children of men . . . Yea, and we may see at the very time when he doth prosper his people . . . *then is the time that they do harden their hearts, and do forget the Lord their God* . . . and this because of their ease, and their exceedingly great prosperity. And thus we see that except the Lord doth chasten his people with many afflictions . . . *they will not remember him.* O how foolish, and how vain, and how evil, and how devilish, and how quick to do iniquity, and how slow to do good, are the children of men. . . . Yea, how quick to be lifted up in pride; yea, how quick to boast, and do all manner of that which is inequity; and *how slow are they to remember* the Lord their God, and to give ear unto his counsels, yea, how slow to walk in wisdom's path! (Hel. 12:1-5).

Appreciation for divine blessings, even major blessings like deliverance from captivity, does not come naturally to fallen man.

> Our fathers understood not thy wonders in Egypt; *they remembered not the multitude of thy mercies;* but provoked him at the sea, even at the Red Sea (Ps. 106:7).

Ye do not remember the Lord your God in the things with which he hath blessed you, but ye do always remember your riches, not to thank the Lord your God for them; yea, your hearts are not drawn out unto the Lord, but they do swell with great pride, unto boasting, and unto great swelling, envyings, strifes, malice, persecutions and murders, and all manner of iniquities (Hel. 13:22).

Lest we arrogantly dismiss such indictments as pertaining to the wicked but not to us, the Book of Mormon teaches that this natural-man trait of slow memory and quick forgetfulness plagues even the most righteous of disciples, such as the brother of Jared, one of the few who possessed sufficient faith to pierce the veil and see the Lord face-to-face. Yet even he, like the rest of us, was distracted by the press of mortal affairs, and so "for the space of three hours did the Lord talk with the brother of Jared, and chastened him *because he remembered not* to call upon the name of the Lord" (Ether 2:14).

Because of the constant barrage of contradictory things presented by this imperfect world, our faith and testimonies are in constant jeopardy and require continual renewal. When we forget the hand of the Lord in our lives and the lives of others, our faith and testimony can quickly wither and shrivel away.

And it came to pass that . . . the people began to forget those signs and wonders which they had heard, and began to be less and less astonished at a sign or a wonder from heaven, insomuch that they began to be hard in their hearts, and blind in their minds, and began to disbelieve all which they had heard and seen (3 Ne. 2:1).

DON'T BE "DOUBLE MINDED"

Another major challenge presented by our fallen minds is keeping them set once we've made a choice. Because our minds are imperfect, we make promises and covenants, then often find ourselves wavering in the face of the first opposition or temptation. When I worked in aerospace, we built autopilots for jet fighters and commercial airliners. We were required to coat the threads of every screw, nut, and bolt with a gooey liquid called "locktite" before they were placed into the assemblies. We coated the threads to keep the parts set where placed and to prevent them from vibrating loose as the planes flew.

The Lord works hard to help us get our minds set so that we don't "vibrate loose" as we pass through the trials of life that would make us waver in our decisions. "Now we beseech you, brethren . . . that ye be not soon *shaken in mind*, or be troubled" but let "every man be *fully persuaded* in his own mind" (2 Thess. 2:2; Rom. 14:5). "Remain *steadfast* in your minds," the Lord counseled, "and "let not your minds turn back" (D&C 84:61; 67:14). We've all experienced the lack of confidence that results when we let our minds waver in their commitment or actually abandon promises and "turn back." "O all ye that are pure in heart, lift up your heads and receive the pleasing word of God, and feast upon his love; for ye may, *if your minds are firm*, forever" (Jacob 3:2).

Setting our minds *firmly* is a matter of mental and emotional commitment, an indication of how seriously we regard this probation. If we're repenting of mistakes, for example, he asks his children to "turn to the Lord their God, with *all* their hearts and with all their might, mind, and strength" (D&C 98:47). If we agree to provide service to him and his kingdom, we're asked to "serve him with *all diligence of mind*" (Mosiah 7:33) and to "serve God with *all* [our] mights, minds and strength" (D&C 20:31). Thus he pleads, "Thou shalt love the Lord thy God with *all thy heart* [the center of our emotions], and with *all thy soul* [the combination of spirit and physical body], and with all thy strength [the mental and emotional energy we devote to his cause], and with *all thy mind* [the center of our thoughts]" (Luke 10:27; see also Matt. 22:37; Mark 12:30; D&C 59:5).

UNDERSTANDING THE MIND

I recently replaced my ancient computer with a new system. Computers are supposed to save us time, but because of the complexity and unfamiliarity of the new software, it took quite a while for me to learn how to use the new equipment's advanced features. Now that I understand it, my processing time is reduced and my capabilities are greatly expanded. But until I understood how this thing works, trying to get the results I wanted was frustrating and time-consuming.

It's the same with the mind, which affects everything we feel, value, and choose to act on. As we learn to *think* better, we learn to *feel* better and to *act* better. Likewise, as we improve control over our emotions, we learn to have better thoughts. Each cycle reinforces the other. The higher our thoughts, the higher our feelings and actions. The lower, or more animalistic, our thoughts, the lower and more animalistic will be our behavior.

By learning *how* the mind and brain function, we can use them to our advantage, rather than misusing them and reaping the consequent difficulties. For example, why do Boy Scouts, policemen, and military personnel wear uniforms? Why do we wear Sunday clothes to church instead of grubbies? Why does the schoolteacher demand that we sit up straight in our desks? It's because you cannot separate the physical from the mental. A slouchy posture influences a slouchy attitude, just as the kind of clothes we wear on the *outside* of our bodies affects our *inner* state of mind. The reason we start church meetings with prayer and song is to redirect our minds away from preoccupation with the multitude of scattered thoughts and activities that preceded our arrival and unite us in a common mind-set of worship and attention to spiritual things. At times like this, our minds can be filled with spiritual resolve and determination, burning with a sincere desire and intention to attain the image of Christ in our thoughts and actions. At other times, our minds can be filled with nothing but thoughts of entertainment, physical pleasure, and gratification. The following chart compares some of the specific positive and negative states of mind which concern the Lord.

SAMPLE STATES OF MIND IDENTIFIED IN SCRIPTURE

Mental attributes to be avoided	Mental attributes to be sought after
Being "carnally minded" (Rom. 8:6-7; Alma 30:53; D&C 67:12)	Being "spiritually minded" (Rom. 8:6)
Having a "wicked" and "reprobate mind" (Prov. 21:7; Rom. 1:28)	Having a "firm mind" and a "firmness of mind" (Moro. 7:30; Jacob 3:1)
Having a "fleshy" and "natural" mind (Col. 2:18; D&C 67:10)	Having a "diligence of mind" (Mosiah 7:33)
Perpetuating "sorrow and grief of mind" (Deut. 28:65; Gen. 26:35)	Being "steadfast in your mind" (D&C 84:61)

Being "weary and troubled in mind" (D&C 84:80; Alma 22:3)

Having a "vanity of mind" or a "a mind hardened in pride" (Eph. 4:17; Dan. 5:20)

Having a "doubtful mind" (Luke 12:29)

Being "double minded" (James 1:8; 4:8)

Having "idle thoughts" and "lightmindedness" (D&C 88:69; 88:121)

Regarding spirituality "lightly" (D&C 84:54)

Having a "forgetful mind" (Alma 46:8)

Having "a frenzied mind" (Alma 30:16)

Being "shaken and troubled in mind" (2 Thess. 2:2)

Having a "blindness of mind" (Ether 4:15)

Entertaining evil imaginations and filthy dreams (Gen. 6:5; Jude 1:8)

Having a "willing mind" (1 Chr. 28:9; 2 Cor. 8:12; D&C 64:34)

Having a "ready mind" and a "readiness of mind"(1 Pet. 5:2; Acts 17:11)

Being in our "right mind" (Mark 5:15)

Having a "sound mind" (2 Tim. 1:2)

Having a "humility of mind" (Acts 20:19)

Having a "lowliness of mind" (Phil. 2:3)

Having thoughts of virtue (D&C 121:45)

Having a "sober mind" and being "sober minded" (Morm. 1:15; Titus 2:6)

These scriptures lead us to the law of mental function, which is extremely important to spirituality. The Bible states it this way: "As he thinketh in his heart, so is he" (Prov. 23:7). There's another way to express this important law: "*What holds my attention holds me.*" Expressing the law of mental function this way is helpful because we all have difficulty controlling our *thoughts*, but all of us can learn to control what we allow to hold our focus of *attention*.

Someone has said, "The difference between a man of weakness and one of power lies not in the strength of personal will, but in that *focus of consciousness* which represents their states of knowledge." *Whatever you give*

your attention to is the thing that governs your life. Attention is the key. Your
free will lies in the directing of your attention. *Whatever you steadfastly
direct your attention to, will come into your life and dominate it* (Emmet Fox,
The Sermon on The Mount [New York: Harper & Row, 1966], p. 109).

The best way I know to explain the relationship between our
thoughts and our focus of attention is to compare the resulting "mind-
set" or "frame of mind" to the function of an electromagnet. A bar of
metal, in its dormant state, is inert and inactive. It has no power to
attract or repel another piece of metal. It's neutral. If it were a mind,
we would say that it was blank with no particular mood or mind-set.
But when an electrical current flows through that same bar of metal,
the molecular structure is reorganized so that it becomes magnetized
with positive and negative poles. That lifeless, inert piece of metal is
charged with magnetic *power* because it has been *polarized.* Now it will
either attract or repel other magnets and pieces of metal.

Similarly, our minds also become *polarized* by the ideas, values,
and concerns we flow through our thoughts and hold in the focus of
our attention. We can flow positive, spiritual, nourishing, and uplifting thoughts, or negative, base, carnal, and degrading thoughts. Two
of the most famous slogans in the computer world are "Input =
Output," and "Garbage In = Garbage Out." It's the same with our
computer minds. Focus on positive, uplifting, and encouraging
thoughts and your mind will be charged with positive power. It will
be "polarized" or "magnetized," as it were, to attract spiritual
thoughts, people, and circumstances that will advance your progress.

On the other hand, when we focus on negative, self-defeating
thoughts, our mind will be "polarized" to act out those commands, for
that's exactly how our minds interpret the things we hold within the
focus of our attention: as the desires, instructions, and *very orders* we wish
them to fulfil. For example, if you're always saying something like "I just
can't remember names," the next time you meet someone, your mind
will, in effect, say something like, "Okay, now I'm going to distract your
attention so that you won't remember this person's name, *because that's
what you told me to do.*" Have you ever said, "I just can't lose weight.
Everything I eat goes to fat"? What was the result of this unintended
instruction to your mind? *What holds your attention holds you.*

A bad habit or an addiction is the result of a mind that's had so many negative or evil desires and thoughts run through it that it's become saturated, "polarized" or "magnetized" so as to be repelled by righteous things and attracted to the very things we want to resist. An addicted mind is like a giant, powerful magnet that *attracts* evil thoughts, circumstances, and temptations. Consequently, what may have begun as a normal human weakness can, with repeated thoughts and intense feelings, cause the negative, evil polarity of the mind to grow stronger and stronger until it becomes a monster we can no longer control. In this state of mind, even the slightest thought or temptation can unleash the huge, accumulated power of this magnetic, mental monster so that over and over we're pulled into the sin we hate.

The power of an electromagnet depends less on the metal than on the amount of current running through it. The greater the flow of electrical current, the more powerful the magnet becomes. Our mind-set works the same way. The strength of a mental addiction is proportionate to the frequency and intensity of the thoughts and feelings we run through the mind and hold in the focus of our attention.

There's one important difference between the magnetism and polarity of the metal and the magnetism and polarity of the mind. As soon as the current is removed from the bar of metal, it returns to its former state of being: lifeless and inert. The magnetism and power have vanished. But the mind isn't like that. The polarizing, magnetizing, addicting power of the intellectual and emotional currents we run through our brain cells is *cumulative*. Every thought and feeling adds to those that went before. Every time we give in and indulge the unworthy thoughts and desires that Satan (or our carnal nature) is suggesting, the strength of the mind's polarity is increased. It accumulates and grows stronger, thought by thought, choice by choice, sin by sin, day by day.

But the law of accumulation is good news, because it can work *for* us just as easily as it works *against* us. As Heber J. Grant was so fond of saying, "That which we persist in doing becomes easier for us to do; not that the nature of the thing itself is changed, but that our power to do is increased" *(Conference Report*, Apr. 1901, p. 63). Understanding that it was the process of accumulation and mental polarization that got us into slavery in the first place will help to free us. Instead of expecting to erase old habits with the simple utterance of a new resolve, we use the power of accumulation for our good, by "reversing the current" as we replace the bad habits and thought patterns with better ones. "Be not overcome of evil, but overcome evil with good" (Rom. 12:21).

Satan would have us believe that our weaknesses, bad habits, and addictions are permanent. But they need not be. When we *reverse the current* through the bar of metal, the polarity is reversed. The negative pole becomes positive. What once attracted now repels. We can do the same with our mind-set by changing the way we think. Through the power of Jesus Christ, every person willing to pay the price can change his or her thought patterns and focus of attention. This change will reverse the current and repolarize the mind so that it will appreciate and be attracted to spiritual thoughts and values instead of the carnal things that come to us naturally. (We'll discuss this further at the end of this chapter.)

We use our minds for opposites: to *remember* things and to *forget* things. In addition to our chosen thoughts and focus of attention, our *memories* also form a major portion of the mental current running through our minds. Almost half of all the Lord's revelations about our minds and thoughts pertain to the memory. We'll now discuss some specific things he has asked us to *forget* and specific things he has asked us to *remember* so that our minds will be properly polarized.

FORGETTING THE PAST

One of the devil's favorite weapons could be symbolized by a rear-view mirror which holds our attention on the past. Though necessary for any driver wanting to stay aware of traffic behind his moving car, the rear-view mirror would be disastrous if most of that driver's attention was focused there instead of on where he wants to go. As Elder Marvin J. Ashton said, "Where you've *been* is not nearly as important as where you are and where you're *going*" (*This People*, March 1984, p. 27).

Satan works hard to keep us looking in our spiritual rear-view mirrors because he knows that a person can't fix *today* as long as *yesterday* remains the focus of attention. For example, whenever a person resolves to draw closer to the Lord, Satan invades and floods his or her mind with memories of past mistakes in an attempt to discourage the effort to change. When Satan gets us to focus too much on the mistakes of the past, that negative focus unwittingly perpetuates the very thing we are trying to overcome because *what holds your attention holds you.*

The greatest barrier to spiritual progression often isn't the continuation of the sins that got us into trouble, but rather our inability to let go of the memories and accept God's forgiveness, which he promised to grant after our repentance. I doubt if anyone endures more misery and necessary suffering than the person who won't take his attention and memory off the past, who insists on going through life with that mental rear-view mirror stuck in front of his eyes, always looking back, always digging up the mistakes of the past, continually criticizing and condemning himself for his failures. Surely this does not please the Lord.

If we're constantly looking backward and condemning ourselves, this misuse of our minds can make us our own worst enemy. By that negative focus, by the very act of dwelling on, analyzing, regretting, pondering, and forcing our whole life to revolve around a problem, we can actually increase its strength and unwittingly perpetuate the very thing we were trying to overcome.

This preoccupation with past mistakes locks guilt and resentment *in* and forgiveness *out!* We think we're being crushed by the demands of justice, when it's only our failure to take advantage of the atonement. As Elder Richard G. Scott said, "Can't you see that to continue to suffer for sins, when there has been proper repentance and forgiveness of the Lord, is not prompted by the Savior, but by the master of deceit, whose goal has always been to bind and enslave the children of our Father in Heaven?" (*Ensign,* May 1986, pp. 11-12).

Alma deservedly suffered terrible memories of his past. He said, "And now, for three days and for three nights was I racked, even with the pains of a damned soul . . . I was harrowed up by the memory of my many sins" (Alma 36:16-17). Most of us can relate to those feelings, but it is the very point of defeat at which Satan seeks to keep us chained. Alma's conversion illustrates the path to freedom from painful memories and guilt. He said that as he placed his faith in Jesus Christ and his atoning blood, his memories were cleansed and healed: "*I could remember my pains no more*; yea, I was harrowed up by the memory of my sins no more. And oh, what joy, and what marvelous light I did behold; yea, my soul was filled with joy as exceeding as was my pain!" (Alma 36:19-20).

True mental and emotional healing requires us to surrender our memories of past mistakes to the Lord just as Alma did, and then, being willing to let go of them, to press forward and make the future what God would wish it to be. Paul realized that he had not yet attained the perfection the Lord commanded us to seek, but he knew it was important to let go of his past mistakes and focus on the goals of the future. "Brethren, I count not myself to have apprehended: but this one thing I do, *forgetting those things which are behind*, and reaching forth unto those things which are before, *I press toward the mark* for the prize of the high calling of God in Christ Jesus" (Philip. 3:13-14).

If we want to win our own battles with memory, we must do the same. "Wherefore, *ye must press forward* with a steadfastness in Christ, having a perfect brightness of hope" (2 Ne. 31:20). If we refuse to let go and forget, we can become so trapped by the guilt and shame we feel for past mistakes, or by resentments for wrongs we have received from others, that we can't believe or receive the forgiveness and healing Christ is trying to give us when we repent. As Dr. Lloyd Ogilvie said, "We mortgage the future based on what happened in the past" (Quoted by H. Norman Wright, *Making Peace with Your Past* [Old Tappan, New Jersey: Fleming H. Revell Co., 1985], p. 40).

Please don't misunderstand. Of course repentance involves looking back. We do that every Sunday during the sacrament. "Thus saith the Lord of hosts; *Consider* your ways" (Hag. 1:7). "*Ponder* the path of thy feet, and let all thy ways be established" (Prov. 4:26). One of the most important things he wants us to think about is consequences so that as our minds make choices, they learn to look past the immediacy of present circumstances to the long-range effects of those choices. Look past today and yesterday, the Lord pleads, and walk "uprightly before me, considering the *end* of your salvation" (D&C 46:7). "O that they were wise . . . that they would consider their latter end!" (Deut. 32:29).

But we must never confuse self-evaluation with self-deprecation, or mistakenly substitute self-punishment for throwing ourselves upon the mercy and forgiveness Christ is so anxious to share with us. Our preoccupation with the past is often like setting up a mental video that never stops. Over and over it replays painful memories, dragging the person lower and lower into despair and self-loathing. Elder Neal A. Maxwell has counseled: "Some of us who would not chastise a neighbor for his frailties have a field day with our own. We should, of course, learn from our mistakes, but *without forever studying the instant replays* as if these were the game of life itself" (*Ensign*, Nov. 1976, pp. 13-14). The loving forgiveness of our Savior and his healing of our memory can turn that video replay off. The Savior's incredible sacrifice can stop the cycle of self-condemnation and resentment if we only accept it and stop punishing ourselves. The Savior is anxious to forgive us, to set us free and heal us of our past, but he won't force it on us. We must first open our hearts and let go of the past. We must give him *permission* to enter our hearts and memory banks and remove the pain we can't remove by ourselves. As Elder Richard G. Scott said:

> Yet there are others who cannot forgive themselves for past transgressions, even knowing the Lord has forgiven them. Somehow they feel compelled to continually condemn themselves and to *suffer by frequently recalling the details of past mistakes.* . . .
> Satan would encourage you to continue to relive the details of past mistakes, knowing that such thoughts make progress, growth, and service difficult to attain. . . . When memory of prior mistakes encroaches upon your mind, turn your thoughts to Jesus Christ, to the miracle of forgiveness and renewal that comes though him. Then your suffering will be replaced by joy, gratitude, and thanksgiving for his love *(Ensign,* May 1986, pp. 11-12).

Many of us have a terrible time forgetting our own sins or the wrongs others have done to us. They fester and lurk in our memories even when we've repented or forgiven the offender to the best of our ability. Years later, if we repeat the error or the offender repeats his mistake, the old wounds appear instantly. When we read that God doesn't remember our forgiven sins, it's hard for us to believe him. We can learn the divine principle of proper forgetting and cleansing of *our* memories by looking at how God deals with *his* memories of the past which no longer serve a purpose.

The Lord said, "I will be merciful to their unrighteousness, and their sins and their iniquities *will I remember no more*" (Heb. 8:12).

"Behold," said the Savior, "he who has repented of his sins, the same is forgiven, *and I, the Lord, remember them no more*" (D&C 58:42). Centuries earlier, he said, "I will forgive their iniquity," and then stressed: "and *I will remember their sin no more*" (Jer. 31:34). God has perfect knowledge; nothing throughout the universe escapes his notice. So how can a God who knows everything truly forget our sins?

As we struggle to let go of our own memories, I think it's important to note that God does not say he *forgets* the sins. What he says is that "I am he that blotteth out thy transgressions *for mine own sake,* and *will not remember* thy sins" (Isa. 43:25). When he says he *chooses* to remember them no more *"for mine own sake,"* he's teaching us that our minds are too sacred to clutter and poison by keeping score of past mistakes. He is the great I Am, the God of the present. He doesn't live in the past, nor should we. When a mistake has been made but the lesson learned, repentance and forgiveness completed, it's time to dismiss it and move on to the next lesson. If our perfect God chooses not to remember certain things, so should we.

The past can never hurt us once we decide to let it go and allow it to truly *be* our past. Please remember that no matter what mistakes lie in your past, your future is just as spotless as the prophet's. With the Lord's help, you can put anything there you choose.

> Prepare your souls for that glorious day when justice shall be administered unto the righteous, even the day of judgment, that ye may not shrink with awful fear; *that ye may not remember your awful guilt in perfectness,* and be constrained to exclaim: Holy, holy are thy judgments, O Lord God Almighty—but I know my guilt; I transgressed thy law, and my transgressions are mine; and the devil hath obtained me, and I am a prey to his awful misery (2 Ne. 9:46).

USING OUR MINDS TO REMEMBER CHRIST

"Can a maid forget her ornaments, or a bride her attire? Yet *my people have forgotten me days without number*" (Jer. 2:32).

Many who read this book were alive when the first man stepped on the moon. Many of us may live to see the day when mankind will explore Mars. It would be difficult to imagine astronauts landing on Mars and then forgetting about their assigned mission—forgetting to report to NASA, who sent them there, and just whiling away their time in idle pursuits. You and I have been sent to this planet on a mission of far greater importance by the creator of the universe. Yet, because we're fallen, many of us become so preoccupied with daily affairs that we do "while away" our time frivolously, forgetting the God who sent us here. "Beware that thou forget not the Lord thy God" is a frequent command in scripture. (See Deut. 6:12; 8:11-14.)

In spite of the Lord's efforts to keep us in remembrance of him, the scriptures repeatedly demonstrate "how quick the children of men do forget the Lord their God" (Alma 46:8). Many do "not like to retain God in their knowledge," and who are "slow to remember the Lord their God." (See Romans 1:28 and Mosiah 13:29.) Isaiah lamented the mental stupor of his people when he said "they *regard not* the work of the Lord, *neither consider* the operation of his hands" (Isa. 5:12). He even compared their preoccupation with daily affairs and neglect to remember the Lord to their dumb animals. "The ox knoweth his owner, and the ass his master's crib: but Israel doth not know, my people doth not consider" (Isa. 1:3).

Alma said, "Let all thy thoughts be directed unto the Lord," and Christ said "Look unto me in every thought" (Alma 37:36; D&C 6:36). President David O. McKay taught, "What you sincerely in your heart think of Christ will determine what you are, will largely determine what your acts will be" (*Gospel Ideals*, Abridged Edition [Salt Lake City: Improvement Era, 1953], p. 234). But our relationship with Christ depends not only on *what* we think of Christ, but also on *how often* we think of him. The more often we remember Christ, the closer we'll feel to him. Thinking about him often closes the door to other preoccupations and opens the door to our hearts, welcoming him to enter.

Remembering Christ and his and our mission in our daily activities is so important that he's given us the ordinance of the sacrament to be repeated once every seven days. Here we're reminded of the importance of holding the Savior in our focus of daily attention, and we renew our promise to do so. By participating properly in this ordinance, we not only witness to the Father that we're "willing to take upon [us] the *name*" of Christ, but also that we'll try to "*always remember* him and keep his commandments" (D&C 20:77). The problem is that sometimes we're so busy trying to remember and obey a seemingly endless list of commandments and duties, that we miss the power key in the sacrament: remembering *him*; remembering *the person* of Jesus Christ and all that he's done for us. Specifically, the words of the prayer are that we'll "always remember *him* and keep his commandments" (D&C 20:77). The key to keeping the commandments is *"remembering him"* so that "[we] may always have his Spirit to be with [us]" (D&C 20:77). If "remembering" comes first, then obedience follows almost automatically.

President David O. McKay helped us appreciate the seriousness of the sacrament when he said that it's "one of the most sacred ordinances of the Church of Christ" and that "too few communicants attach to this simple though sublime rite the importance and significance that it merits" *(Pathways to Happiness* [Salt Lake City: Bookcraft, 1957], p. 262). Paul referred to the hazards of not taking the memory implications of the sacrament seriously when he said that "he that eateth and drinketh unworthily, eateth and drinketh damnation to himself, not discerning the Lord's body" and that "for this cause many are weak and sickly among you" (1 Cor. 11:29-30). The weakness, sickness, and damnation (or hindrance) to our progress that results from indifferent or unworthy participation in the sacrament can be *physical, spiritual,* or both. If your body has recurring problems with lingering, not specifically definable, sicknesses that make you weak, there's a possibility that the cause could be spiritual rather than some specific germ or disease. The mental and emotional health of the mind, the body, the heart, and the spirit are inseparably connected.

Just as the strength of that electromagnet is dependent on and proportionate to the amount of current passed through the bar of metal, so it is with our minds spiritually. It's like a mathematical ratio. The more often we "remember that there is no other way nor means whereby man can be saved, only through the atoning blood of Jesus Christ," the more we yearn for his Spirit in our lives (Hel. 5:9). The more we learn to hold Christ in the focus of our attention, the closer we feel to him, the more prepared we are to receive his personal fellowship, and the more spiritually minded, "polarized," or "magnetized" our minds will be to attract his Spirit to us.

"WHAT HOLDS YOUR ATTENTION HOLDS YOU"

The reverse of the ratio is also true. The less we focus on Christ and his scriptural promises, the farther away he seems. The less we think about *him*, the more room there is for the problems, temptations, and distractions of the world to crowd into our minds, "polarizing it" to attract things of the world and separating us from the Savior and the fellowship he's seeking to give us. Because the scriptures emphasize that we always reap what we sow, the logical reward for holding Christ in our thoughts is the Savior's promise that "if ye do always remember *me* ye shall have my Spirit to be with you" (3 Ne. 18:11). And isn't that exactly what we all need if we are to know him personally and feel close to him? Elder Dallin H. Oaks emphasized how important the sacrament is to this goal: "The close relationship between partaking of the sacrament and the companionship of the Holy Ghost is explained in the revealed prayer on the sacrament. To have the continuous companionship of the Holy Ghost is the most precious possession we can have in mortality" *(Ensign,* Nov. 1998, p. 38).

King Benjamin verified that if we would be close to Christ, "Ye should remember to retain [his] name written always in your hearts, that ye are not found on the left hand of God, but that ye hear and know the voice by which ye shall be called . . . *For how knoweth a man the master . . .* who is a stranger unto him, and *is far from the thoughts and intents of his heart?* (Mosiah 5:12-13). We can't expect to make a personal friend of someone we hardly ever think about except for a few minutes once a week in church. The rewards for "remembering him always" are incredible, for "thou wilt keep him in perfect peace, whose mind is *stayed* on thee" (Isa. 26:3).

The Lord is so appreciative of those with the spiritual discipline to keep the sacrament covenant that "a book of remembrance was written before him for them that feared the Lord, and that thought upon his name" (3 Ne. 24:16; see also Mal. 3:16). When Christ returns, those whose names are in this record will be singled out for special greeting. "Yea, when thou comest down, and the mountains flow down at thy presence, thou shalt meet him who rejoiceth and worketh righteousness, *who remembereth thee* in thy ways" (D&C 133:44). But "all they who are *not* found written in the book of remembrance shall find none inheritance in that day, but they shall be cut asunder"

(D&C 85:9). "And now behold, I say unto you, my brethren, you that belong to this church, have you sufficiently retained in remembrance . . . ?" (Alma 5:6).

THE EFFECT OF THE SCRIPTURES ON THE MIND

"And your minds in times past have been darkened because of unbelief, and *because you have treated lightly* the things you have received" (D&C 84:54).

The person who doesn't use the scriptures to "polarize" his mind is just as spiritually deprived as a primitive in the deepest jungle who doesn't even know the scriptures exist. Jesus challenged us to *"search* the scriptures," to *"study* my word which hath gone forth among the children of men," to "go ye unto your homes, and *ponder* upon the things which I have said," and to *"treasure* up in [our] minds continually the words of life." (See John 5:39; D&C 11:22; 3 Ne. 17:3; and D&C 84:85.) Yet it seems that our desire to search and ponder the scriptures is seldom strong enough to persuade us to set aside the necessary time.

I've worked with hundreds of people who wanted to improve their spiritual lives. I've never seen one of them succeed who did not first fall in love with the scriptures. I'm sorry if that disappoints you. But from my own personal experience with *using* the scriptures and from my experience with *neglecting* them, it's my conviction that those who believe they can achieve holiness and spirituality without being close to the scriptures are deceiving themselves. I know that Satan would be very pleased with that delusion. Here's a quote that might help you to believe me. Church Patriarch Eldred G. Smith said: "The Lord made no promise to those who try to go it alone. As soon as you think you can lick the devil alone, on your own, without the Lord's help, you have lost the battle before you start" (*Ensign*, Dec. 1971, p. 46).

In my opinion, there's nothing more effective people can do toward gaining happiness and victory over their carnal natures and improper thoughts than to read and ponder the scriptures daily. Many people struggle with this issue and feel guilty because of their neglect. There's so much competition for our time that daily reading of the scriptures seems to require reasons that are higher than duty. Here are four other reasons that can help.

THE SCRIPTURES AID OUR MEMORY

With all the distractions competing for our attention, it would be very difficult to honor our sacrament covenant to remember the Savior in our daily affairs without the help of the scriptures. One of the reasons for *daily* scripture study is to *"remember* the words of him who is the life and light of the world, your Redeemer, your Lord and your God" (D&C 10:70). Unfortunately, it's part of our natural, mortal pride to believe we don't need to read the scriptures daily because we can remember the commandments without daily input. Of course it's not true, because "it were *not possible* that [even] Lehi, *could have remembered* all these things, to have taught them to his children, except it were for the help of these plates [scriptures]" (Mosiah 1:4). "And now, it has hitherto been wisdom in God that these [scriptures] should be preserved; for behold, *they have enlarged the memory* of this people, yea, and convinced many of the error of their ways" (Alma 37:8).

We'd more effectively honor our sacrament covenant to remember Christ if we were with him *physically* each day. That cannot be, but we can be with him *mentally* and *emotionally* as much as we want by visiting with him through the scriptures. You can't separate Christ from the scriptures. Christ is "the Word." (See John 1:1-4.) Because the *will* of the Lord and the *word* of the Lord are one and the same, the more time we spend reading, searching, and pondering his scriptural *words*, the better we'll know him and his will for our lives. In addition, we'll be more likely to recognize his hand in our daily lives. Most importantly, we'll feel love *for* him and *from* him more strongly than ever.

Jesus has lovingly pleaded, *"Remember* the words of him who is the life and light of the world, your Redeemer, your Lord and your God" (D&C 10:70). The Savior feels so strongly about our remembering the scriptures that he actually equates our feelings toward *them* to our feelings toward *him.* "If a man love *me,* he will keep my *words"* (John 14:23). "Keeping" his words means more than obedience to his commandments. It also means remembering them and keeping them focused in our thoughts. How could we possibly claim to love *him* and yet be indifferent to his *word?* Would we call him a liar by claiming to love him while we treat his words with casual indifference, putting forth little effort to learn and remember them? "For where your treasure is, there will your heart be also" (Matt. 6:21). "I will delight myself in thy statutes: I will not forget thy word. I will never forget thy precepts: for with them thou has quickened me" (Ps. 119:16; 93).

THE SCRIPTURES HELP US TO KNOW GOD

> In order to feel comfortable with myself, I must get to know *myself* as God reveals me in the scriptures. In order to feel comfortable with *God,* I must get to know God as he reveals *himself* in the scriptures (Don Baker, Acceptance [Portland, Oregon: Multnomah Press, 1985], p. 122).

Simply *remembering* Christ isn't enough. To become like him, we must know him. Neither the Father nor the Son will ever become familiar to us until we're familiar with their recorded words. How could we possibly know the Lord or his love for us if we don't know his word as revealed in scripture? As the Lord said, "Whoso receiveth not my voice is not acquainted with my voice, and is not of me" (D&C 84:52). One of the main reasons we're asked to search and ponder the scriptures is to help us know our Savior and Heavenly Father more intimately, as well as what we have the right to expect from them. When the Lord places the record of his dealings with specific individuals or groups of people in the scriptures, he does it to show that you and I have the right and privilege to lay claim on any principle or promise we find recorded there. For being no respecter of persons, he has promised "that what I say unto one I say unto all" (D&C 61:18; see also 61:36; 82:5). How can we rely on the Lord for strength and guidance in time of need if we don't know what he has promised in his scriptures to do for us? "The scriptures teach us how the Father has related to his children in the past; more important, they teach us how he will relate to *us* if we are worthy and if we seek him" (Mollie H. Sorensen, *Ensign*, March 1985, p. 27).

We can trust this principle because God never changes. The scriptures are filled with the assurance of God's constancy. "From eternity to eternity he is the same" (D&C 76:4). "For do we not read that God is the same yesterday, today, and forever, and in him there is no variableness *neither shadow of changing*?" (Morm. 9:9). Because God is perfectly stable, perfectly constant and consistent, and most importantly, perfectly predictable, we have every right to expect him to respond to us in the same way he has responded to others in the past.

Without this assurance that we can depend on the constancy of God's words and feelings, the constancy of his attitude, love, and goodwill, his patience and forgiveness, without the assurance that these and all his other attributes we find in the scriptures will remain the same, how could we ever dare to trust him or place our faith and hope in his promises? Time and again it has been proven that the intimacy of love we feel *for* the Lord and *from* the Lord will be increased or decreased in proportion to our study or neglect of the scriptures.

THE SCRIPTURES CREATE FAITH

"But without faith it is impossible to please him" (Heb. 11:6).

It's impossible to please God without faith because we can't conquer the natural man and return to his presence without it. "They shall have faith in me," he declared, "or they can in nowise be saved" (D&C 33:12). "Remember," he pleaded, "that without faith you can do nothing," because "I am a God of miracles; and . . . I work not among the children of men save it be according to their faith" (D&C 8:10; 2 Ne. 27:23). It isn't simply *faith*, however, that prepares us for life in the celestial kingdom. It's *faith in Jesus Christ.* Just as sunlight passing through a magnifying glass gains no power until it's focused, so it is with faith. Only when we focus our faith on Christ do we gain the hope and power to take us back to the Father. But how do we obtain or create this essential faith?

We know that if we want to harvest vegetables from our garden, we must first pay the price of cultivating the soil, planting the seeds, and providing water and nourishment to the soil during the growing months. It's the same with developing faith, and it's the same with building our bank accounts. It isn't a very pleasant experience to have a financial emergency and find the checkbook empty or insufficient. Nor is it a pleasing experience to face a spiritual emergency and find our faith weak and insufficient. I think we'd all agree that the more money we have on deposit, the more secure and safe we feel financially. But there's an inescapable principle that we can't withdraw something we haven't first deposited. If we want to have money available when our bills come due, we have to pay the price of making the deposits in *advance* of the need. And it's exactly the same with faith and spirituality. Building the strength of belief, faith, and trust that will sustain us during times of trial and temptation requires the planting of many spiritual seeds and the repeated deposits of scriptural principles and promises in our minds *in advance* of the need.

Creating faith in Christ is related to the inescapable mental function of the electromagnet. The way the mind decides what we *really* believe, trust, and expect to happen is by majority rule, by the frequency of what we "flow through it" or tell it most often. It was Paul who taught us how to *increase our faith* by reading scripture. "Faith *cometh*," he said, "by *hearing* . . . the word of God" (Rom. 10:17). In today's world of printed scriptures, "hearing" would include reading. One reason the brethren urge us to spend time in scripture study every day is not just to obtain new *knowledge*, but to provide the spiritual current that will polarize our minds with faith and fix the belief and trust in our minds. Faith "cometh," or grows and develops, by reading and thinking about the words of God over and over. Every time we run a scriptural promise or principle through our minds, we're literally planting spiritual seeds and making mental deposits the holy Spirit can draw on when we're in need of sustaining faith and belief.

Alma gave us a formula for increasing faith and spiritual strength. "As much as ye shall put your trust in God," he said, "even so much ye shall be delivered out of your trials, and your troubles, and your afflictions" (Alma 38:5). This means the *more time* we spend in the scriptures and keeping the promises focused in our minds, the more victorious and holy we'll be because our faith will be strong and it will be healthy and growing. On the other hand, the *less time* and effort we invest in studying and digesting and remembering his words, the more battles we'll lose and the more discouraged we'll feel, because our faith will be withering from lack of spiritual nourishment.

If we use even a small part of each day to consistently plant the seeds of faith by making deposits of gospel truths and divine principles and promises in our minds and hearts, then, when the times of discouragement and temptation and adversity come, we'll have spiritual resources to draw on and we'll be successful in meeting our challenges. But if we're not willing to read the scriptures daily to make these deposits and build a spiritual reserve, how can we expect to reap the harvest of strong faith, belief, and trust when the times come for withdrawal?

It would be wonderful if we could go to the bank today and deposit an extra $10,000. But isn't it true that for most of us, the only way we'll ever have that $10,000 in the bank will be to accumulate a lot of small deposits over time? It's the same in building faith. The Lord's goal is to help us *internalize* his words so that they will provide automatic support for our spiritual needs. "I will put my laws into their mind," he said "and write them in their hearts" (Heb. 8:10). It isn't reasonable to expect to reverse years of negative programming with just a few recitations of a new resolve or an occasional glance at scripture. If we really want to erase doubts and reverse our mental polarity, we have to input and internalize the scriptural deposits steadily, daily, over the weeks, months, and years.

The instant you turn off the current to an electromagnet, it loses its power. Neglecting daily scriptures is turning off the spiritual current. It's like posting a vacancy sign that invites Satan to take over and make the mental and emotional deposits for us. He is surely ready and eager to do just that.

We deny ourselves many blessings when we regard the study of scripture as a difficult and unpleasant task. If this has been a problem for you in the past, please know that you can change. Even if you have to force yourself to study at the beginning, before long, the mind and spirit will recognize and rejoice with the new input. It will begin to feel good. Peter said that we should, "as newborn babes, desire the sincere milk of the word, that ye may grow thereby" (1 Pet. 2:2). Everyone smiles at the eager enthusiasm an infant shows for its feeding, but how many of us apply that kind of enthusiasm to the study of scripture? One sign of true discipleship is that we "*enjoy* the words of eternal life in this world" (Moses 6:59). The more time we spend in the scriptures, the more we'll find ourselves craving time with the Lord instead of with our sins.

THE SCRIPTURES PROVIDE ESSENTIAL NOURISHMENT

If you were on a space trip to a distant planet and your life-support system depended on a complex organic computer system, is there any way you would neglect it or take it for granted? Would a day go by that you didn't check it, nourish it, calibrate it? Well we've already made that long journey. This earth is our "distant planet." If we're to survive the hazards here and make a safe return to our Heavenly Father, our spiritual life depends on the proper use of the computer system our Heavenly Father has provided: the combination of our physical brain and sacred, spirit mind.

We arise each morning and fuel our bodies with breakfast nourishment. We refuel it sometime in mid-day and again in the evening. If we don't, our stomachs protest and remind us of the need for more food. The mind and spirit also need nourishment. That's why we're admonished to read the scriptures daily. Just as the physical body grows weak without nourishment, so does the mind and spirit.

Although the physical brain weighs only about three pounds, a very small part of the entire body mass, it uses about 20% of the body's blood and energy, and 20% of all the nutrition we feed our bodies. This is remarkable when we realize the brain has no energy-hungry muscles. The brain has so many things to do to keep our body running properly that its need for fuel and its consumption of energy is more or less constant whether we're concentrating hard on something or just relaxing or sleeping. The mind and spirit consume mental and emotional energy as well. It's been said that wars are won or lost by the food supplied to the soldiers because of the stamina and energy required to fight their battles. It's the same with our spiritual battles. Moroni spoke of the church's members needing to be "*nourished* by the good word of God, *to keep them in the right way*, to keep them continually watchful unto prayer, relying alone upon the merits of Christ" (Moro. 6:4).

To help us appreciate our continual need for daily nourishment from the scriptures, Christ compared our relationship to *him* to that of a vine and a branch. He said, "I am the vine, ye are the branches" and just "as the branch cannot bear fruit of itself, except it abide in the vine; *no more can ye*, except ye abide in me." He further explained, "He that abideth in me, and I in him, the same bringeth forth much fruit: for *without me ye can do nothing*." But "if a man abide not in me, he is cast forth as a branch, *and is withered*; and men gather them, and cast them into the fire, and they are burned." (See John 15:1-5.) Have you ever felt undernourished and "withered"? If your life is filled with thoughts of discouragement, hopelessness, despair, and weariness, it's your undernourished spirit crying out for help, telling you in the only way it can that it needs nourishment.

We'd never expect our automobiles or trucks to move us about or carry our loads without the proper supply of fuel. How foolish to expect our minds and spirits to carry our burdens without the proper supply of daily scriptural nourishment! I once had a customer who tried to pay for a $20.00 purchase with his credit card. It was refused by the processing computer. I said, "I'm sorry, sir, but your credit charge has been declined." He said, "Try it for $18.00." I did so; it was accepted, and he paid the remaining $2.00 in cash. I was greatly distressed by this man's situation. What a terrible way to live one's financial life, charged with deficit and obligation so close to the allowed maximums! But many of us live our spiritual lives in deficit, trying to "run on empty," as it were, never taking the time to feed and nourish our minds and spirits. And then we wonder why we're so irritable, depressed, and overwhelmed by each little adversity that comes along.

Athletes around the world are busy working out as they prepare their mental and physical fitness for the next Olympics. You and I are engaged in a Celestial Olympics. We're supposed to be using this mortal probation to prepare ourselves spiritually to become the kind of people who deserve to be in celestial perfection and will feel good about being there. The purpose of the gospel is to develop us into the spiritual champions who deserve and reap the rewards of an eternal joy and glory. That puts a mere "gold medal" to shame. But we can't coast our way into the celestial kingdom, any more than an athlete could coast his way into an Olympic gold medal. We have to pay the price of spiritual exercise and training, just the same as the athlete has to pay the price of physical preparation.

While there are many elements to developing physical fitness, I believe you would agree that the foundation would have to be a good nutritional program. What good would it do to try to increase our strength or speed if we're not giving our bodies the fuel required for good physical health and energy?

The same thing is true of spiritual fitness. It takes spiritual energy to face our daily duties; it takes energy to make changes in our lives, to grow and improve, to "endure to the end" faithfully, and that energy can only come from the spiritual nutrition provided by the scriptures.

On my way to work one morning, I stopped to refuel my car. I was distressed to observe a young girl on her way to school who also stopped for refueling. Her inappropriate breakfast was a Coke and a candy bar. I've often thought about her lack of proper nourishment, comparing it to how we starve our spirits and minds when we neglect scriptural nourishment. Nephi advised us to feed our hunger for spiritual nourishment by "*feasting* upon the word of Christ" (2 Ne. 31:20). And the Savior promised his disciples who do literally "*hunger* and *thirst* after righteousness" that they would be blessed by having that yearning filled with the presence and strengthening, sustaining influence of the Holy Ghost. (See 3 Ne. 12:6.) As President Kimball said, "I find that when I get casual in my relationships with divinity and when it seems that no divine ear is listening and no divine voice is speaking, that I am far, far away. If I immerse myself in the scriptures, the distance narrows and the spirituality returns" (*The Teachings of Spencer W. Kimball*, p. 135, as quoted in the *Ensign*, Jan. 1999, p. 4).

THE BLESSING OF A CONSCIENCE

Every part of your body is connected to your brain by a network of more than 90,000 miles of nerves. If stretched end to end, these nerves could cross back and forth across the United States of America more than thirty times! Every moment, countless millions of signals speed through all parts of the body as nerve cells go about their business of signaling the brain with the information necessary to keep the body functioning properly. The average signal or nerve impulse travels through the body at about 165 feet per second, which is more than 100 miles per hour, but some signals can reach the brain at a speed greater than 250 miles per hour. Because of the data perceived by these nerves, it's been estimated that the brain may receive as many as 100 million sensations every second!

Just as the Lord has placed these millions of nerve cells throughout our bodies to detect environmental threats and warn us of conditions that would cause us *physical* pain, he has also placed spiritual "sensors" in our minds called "conscience." Its purpose is to detect and warn us of improper thought patterns and choices that could result in spiritual damage. It does this by causing feelings of guilt and *spiritual* pain. "Now, there was a punishment affixed, and a just law given, *which brought remorse of conscience unto man*" (Alma 42:18). This divinely bestowed process of self-incrimination was described by Paul when he said that "the work of the law [is] written [programmed] in their hearts, their conscience also bearing witness, and *their thoughts the mean while accusing or else excusing one another*" (Rom. 2:15).

> Every person born into the world is endowed with the light of Christ as a free gift (D&C 84:45-48). *By virtue of this endowment all men automatically and intuitively know right from wrong and are encouraged and enticed to do what is right* (Moro. 7:16). The recognizable operation of this Spirit in enlightening the mind and striving to lead men to do right is called *conscience*. It is an *inborn consciousness* or sense of the moral goodness or blameworthiness of one's conduct, intentions, and character, together with an *instinctive feeling* or obligation to do right or be good (Bruce R. McConkie, *Mormon Doctrine* [Salt Lake City: Bookcraft, 1966], pp. 156- 157).

The benefit of this spiritual "warning light" is illustrated in the sinful men who threw the adulterous woman at the feet of the Savior, demanding her condemnation and stoning. When they were confronted by the Savior's intimate knowledge of their own secret sins, they left off their demands for her punishment, and "being *convicted by their own conscience*, went out one by one" (John 8:9). This was Heavenly Father's intent in his gift of conscience: that we might be *convicted privately* and turn from our sins before permanent damage is done.

The power of the conscience to protest our sin can vary all the way from a mild irritation such as "he shall not feel quietness in his belly" (Job 20:20) to the terror felt by Alma, which he described as so powerful that "the very thought of coming into the presence of my God did *rack my soul with inexpressible horror*" (Alma 36:14).

If we try to ignore the conscience's warning signals, they can become a relentless, gnawing mental anguish. Zeezrom felt this anguish because of his persecution of Alma and Amulek, which sins "did harrow up his mind until it did become exceedingly sore, *having no deliverance*; therefore he began to be scorched with a burning heat" and "*his mind . . . was exceedingly sore because of his iniquities*" (Alma 15:3, 5). We don't like the burn we receive when we touch something that is too hot, but imagine how much more damage would be done if we didn't have that instant sensory warning that we should withdraw our touch. Similarly, we don't like the guilty feelings that come when we do things we know we shouldn't do. But imagine how much more damage would be done if we didn't have that warning and continued in the sin until it became addictive and destroyed our spiritual life! There would be far less repentance if we had no pain of conscience to disturb our contentment in sin. "The Lord knoweth the thoughts of man" (Ps. 94:11).

One reason we ignore the warning signals from our consciences and sin in both thought and action is that we think we can do it in secret without being detected. I believe the warning signals that our consciences send to *us* also go to the *Lord,* alerting him to our danger as surely as a loud clanging bell alerts the firemen in a fire station. Police can take a fingerprint and identify a person. Dentists can x-ray our mouths and know exactly the present condition of our teeth as well as what has been done to them in the past. It should be no surprise, then, that God can scan our consciences and know exactly what our thoughts and feelings are, for he "is a discerner of the thoughts and intents of the heart" (Heb. 4:12; see also D&C 33:1). To "discern" something is to detect it, to perceive, recognize it, or comprehend it mentally. To help his children pay more attention to their consciences, the Lord has kindly revealed that he monitors them constantly. "I know your thoughts," he said (3 Ne. 28:6). "I know the things that come into your mind, *every one of them*" (Ezek. 11:5).

The conscience is a very delicate instrument of the spirit mind. It can be easily abused and damaged by neglect. "And he that repents not, from him shall be taken even the light which he has received; for my Spirit shall not always strive with man, saith the Lord of Hosts" (D&C 1:33). With prolonged abuse the conscience can grow numb, as if worn down by repeatedly being ignored. The scriptures speak of those whose deliberately prolonged sin has caused them to degenerate spiritually till they "are past feeling," or have "their conscience seared with a hot iron" until "even their mind and conscience is defiled" (Eph. 4:19; see also 1 Ne. 17:45; Moro. 9:20; 1 Tim. 4:2; and Titus 1:15). Knowing that nothing is secret from Heavenly Father's watchful eye should help us to give greater heed to our consciences. If we don't, the time will come when unrepented sins, evil thoughts, and secret imaginations will be made known in the day of judgment, as "the second angel [shall] sound his trump, and reveal the secret acts of men, and the thoughts and intents of their hearts" (D&C 88:109).

On the other hand, honoring our conscience brings the spiritual gifts of peace and confidence before God. When King Benjamin's people repented and committed their lives to Christ, they not only "received a remission of their sins," but also a "*peace of conscience,* because of their exceeding faith which they had in Jesus Christ" (Mosiah 4:3). The joy that can come from responding to a wounded conscience is often proportional to the pain it causes before our repentance. Thus Alma stated that after his repentance and surrender to Christ, "I could remember my pains no more; yea, I was harrowed up by the memory of my sins no more. And oh, what joy, and what marvelous light I did behold; yea my soul was filled with joy as exceeding as was my pain! Yea, I say unto you, my son, that there could be nothing so exquisite and so bitter as were my pains. Yea, and again I say unto you, my son, that on the other hand, there can be nothing so exquisite and sweet as was my joy" (Alma 36:19-21).

As we seek to perfect our minds, memories, and consciences, we should strive to join Paul, who said, "I have lived in all good conscience before God until this day" and "herein do I exercise myself, to have always a conscience void of offence toward God, and toward men" (Acts 23:1; 24:16). Similarly, King Benjamin said that he spent his life "walking with a clear conscience before God" and that "I can answer a clear conscience before God this day" (Mosiah 2:27, 15). We know this is possible for us as well because the Lord has not only asked us to "let virtue garnish thy thoughts unceasingly" but has promised that as we do so, "then shall thy confidence wax strong in the presence of God" (D&C 121:45). When our thoughts and consciences are clear, we can say, at any time and without hesitation, "Search me, O God, and know my heart: try me, and know my thoughts," for "happy is he that condemneth not himself in that thing which he alloweth" (Ps. 139:23; Rom. 14:22). "Beloved, if our heart condemn us not, then have we confidence toward God" (1 John 3:21).

SPIRITUAL "STROKES"

Brain attacks are called "strokes." They result in the loss of brain function because of a sudden blockage or rupture of a blood vessel to the brain. The resultant loss of muscular control, diminution or loss of sensation or consciousness, dizziness, slurred speech, or other symptoms vary with the extent and severity of the damage to the brain.

Strokes are the leading cause of disability in America and the third largest cause of death. Approximately a half a million people a year suffer these unexpected attacks. However, virtually every accountable person experiences spiritual attacks on their minds, for one of Satan's primary targets is our thoughts. Yes, he tempts us through the desires of our bodies, but the most frequent, intense, and important battles occur within the mind and the thoughts. It's there that we wage hand-to-hand combat with Satan and his determined army of evil spirits as well as with our inherent fallen nature. As President David O. McKay said, "The greatest battle of life is fought within the silent chambers of your own soul" (*Ensign*, May 1980, p. 56).

Except when we're in the temple, "we are surrounded by demons, yea, we are encircled about by the angels of him who hath sought to destroy our souls" (Hel. 13:37). President Wilford Woodruff estimated that Satan would assign about one hundred of these demons to every man and woman in the Church. (See *Journal of Discourses*, vol. 21:125-26.) Satan and his demons attack our minds and thoughts because they know that "to be carnally minded is death; but to be spiritually minded is life and peace" and therefore, whoever controls the *thought* patterns controls the *behavior* patterns (Rom. 8:6). They know that if they can influence us to *think* evil, eventually we will end up *doing* evil. The sin may not come for months, or even years, but sooner or later, we always reap in our physical lives what we first sow in our mental lives. "What holds your attention holds you." Thus, whether we're obedient, valiant, and righteous disciples, or mediocre, lukewarm, or even evil is first determined in the battleground of our thoughts.

The Lord can speak directly to our spirit minds, causing thoughts to appear in our conscious minds. (See D&C 6:23.) Satan can do the same thing. The way he and his evil spirits attack our minds is by following us everywhere we go (except the temple) and whispering things like lies, discouragements, resentments, and temptations in our ears. (See 2 Ne. 28:22.) "There is not a person alive who has not, at some point in his life, had an inappropriate thought enter into his mind, primarily because Satan has the power to put it there" (Ronald A. Dalley, *The New Era*, Aug. 1984, p. 45). "We should be on guard always to resist Satan's advances. *He has power to place thoughts in our minds* and to whisper to us in unspoken impressions to entice us to satisfy our appetites or desires and in various other ways he plays upon our weaknesses and desires" (*Answers to Gospel Questions*, comp. Joseph Fielding Smith, Jr., 5 vols. [Salt Lake City: Deseret Book Co., 1957-66], 3:81).

President Benson compared the battle for our focus of attention to a stage. He said, "From one side of the wings the Lord, who loves you, is trying to put on the stage of your mind that which will bless you. From the other side of the wings the devil, who hates you, is trying to put on the stage of your mind that which will curse you. Usually with our hardly realizing it, he slips into our thoughts" (*Ensign*, Apr. 1984, p. 11). Using the same analogy, Elder Boyd K. Packer asked: "Have you ever noticed that without any real intent on your part . . . a shady little thought may creep into your attention? These delinquent thoughts will try to upstage everybody. If you permit them to go on . . . they will enact for you . . . anything to the limits of your toleration" (*Ensign*, Jan. 1974, p. 28).

Satan's influence has filled the world with filth, calculated to pollute and degrade our thoughts. The Internet, radio and televison programs, popular music, filthy talk shows—the entire range of media is saturated with things that will polarize our minds away from Christ and spiritual values if we allow them to occupy the focus of our attention. We didn't create this wicked environment, and as natural men and women in a fallen world, we can't control what thoughts will suddenly appear on the stage of our thoughts. But we *can* control what remains in our *focus of attention*, and this is the key to controlling our thoughts. President Benson emphasized that neither satanic suggestions nor divinely whispered inspirations can be forced on us. "You are the stage manager," he said. "You are the one who decides which thought will occupy the stage. You are the one who must decide whose thoughts you will entertain" (*Ensign*, April 1984, p. 11).

But what do we do when unworthy thoughts do slip past our defenses and startle us with their presence? What if we're presently under bondage to the habit of evil thoughts?

Traditionally, we've thought of breaking habits and repentance as *stopping* the bad thought, sin, or habit. But when we understand mental "polarization" and how "what holds our attention holds us," we can know that it's easier to change behavior and thought patterns when we have something positive *to do* instead of something negative *not to do*. Instead of ineffectively trying to just *block* bad thoughts, we can reverse our mental polarity by *replacing* them with good thoughts. We can do this by turning to the scriptures, by learning to hold Christ in the focus of our attention, and by changing what we do with our time. *Whatever dominates our time dominates our thoughts*. If you wish to change your thoughts, change what you're doing with one of your most precious assets: time. Elder Boyd K. Packer explained this in an address at BYU on February 21, 1978:

> Do not try merely to discard a bad habit or a bad thought. Replace it. When you try to eliminate a bad habit, if the spot where it used to be is left open it will sneak back and crawl again into that empty space. It grew there; it will struggle to stay there. When you discard it, fill up the spot where it was. Replace it with something good. Replace it with unselfish thoughts, with unselfish acts.
>
> Then, if an evil habit or addiction tries to return, it will have to fight for attention. Sometimes it may win. Bad thoughts often have to be evicted a hundred times, or a thousand. But if they have to be evicted ten thousand times, never surrender to them. You are in charge of you.
>
> I repeat, it is very, very difficult to eliminate a bad habit just by trying to discard it. Replace it.
>
> Read in Matthew chapter 12, verses 43 to 45, the parable of the empty house. There is a message in it for you (*Devotional Speeches of the Year* [Provo, Utah: Brigham Young University Press, 1978], p. 39).

The wonderful thing about our minds is that they can't entertain evil thoughts and spiritual thoughts at the same time. With all the distractions of this mortal world and the constant attacks by Satan on our thoughts, keeping the Lord's Spirit on the stage of our mind is not easy. It takes commitment and perseverance. It's one of the most important efforts we can make in this life. That's why we're asked to read scriptures daily and repeat the ordinance of the sacrament every seven days. To *remember* the Lord is to *keep* his Spirit. To *forget* the Lord is to *lose* his Spirit, and *to lose the Lord's spirit's to lose our battles*, for as Elder Joseph B. Wirthlin said of pornography, for example, "Every *ounce* of pornography and immoral entertainment will cause you to *lose a pound* of spirituality. And it will only take a few ounces of immorality to cause you to lose all of your spiritual strength, for the Lord's spirit will not dwell in an unclean temple" (*The New Era*, May 1988, p. 7). By filling our minds with spiritual thoughts, we place a "no vacancy" sign on our mental stage that keeps Satan's influence out.

Cycling the Savior's scriptural promises through the mind over and over not only replaces the immediate bad thoughts we're fighting, but also changes our mental polarity and replaces the old failure programming and expectations of failure. We need to saturate our brain cells with Christ's promises of deliverance. Then, when temptation or discouragement try to bring us down, instead of being pulled back into the cycles of struggle and failure that have been programmed into our expectations, the spiritual current we've been running through our "electromagnet-minds" will have us charged and polarized. We'll be armed with truths of deliverance that will repel Satan's negative suggestions and draw us into the power of Christ. The temptations and old patterns of thought simply cannot withstand the accumulated power of Christ's words when we internalize them by keeping them in the focus of our daily attention.

The instant you release a rubber band that has been stretched, it snaps back to its inert size. This world is so full of distractions that unless we're constantly flowing the polarization of the scriptures through our minds, they also tend to "snap back" into worldly preoccupations and lose their focus on spirituality. Of course we can't always grab a book of scripture at the moment of temptation or adversity, so another way to repolarize the mind and "snap it back" into a state of faith and spirituality is to sing or even hum a hymn. Music has a mental and emotional power that puts mere words to shame. Sacred music can change one's state of mind and mental focus instantly.

By saturating our minds with thoughts of Christ and thoughts and principles from the scriptures and sacred music, we can train them to snap back into a spiritual mode every time Satan tries to distract them with worldliness. Then, as Paul emphasized, we can have the total victory of "casting down imaginations . . . and bringing into captivity *every thought* to the obedience of Christ" (2 Cor. 10:5). In the last chapter we'll discuss this "mighty change." But if we don't learn to control the thoughts that occupy our minds in this life, we'll one day find ourselves standing before the Lord in great shame because "we will not be found spotless; and *our thoughts will also con-demn us*; and in this awful state *we shall not dare to look up to our God*; and we would fain be glad if we could command the rocks and the mountains to fall upon us to hide us from his presence" (Alma 12:14).

Thankfully, the Lord has promised that if we will keep our mind-temple clean and occupy it with wholesome, spiritual thoughts, and if we "let virtue garnish [our] thoughts unceasingly; then shall [our] con-fidence wax strong in the presence of God" and we'll rejoice to be in his presence. (See D&C 121:45.) "Let . . . the meditation of my heart be acceptable in thy sight, O Lord, my strength, and my redeemer" (Ps. 19:14).

WHAT WE KNOW ABOUT GOD'S THOUGHTS

One of the most compelling reasons to reverence our minds and use them as God has asked us to is the way he uses his mind on our behalf. The scriptures repeatedly emphasize that even with all the things required of him to operate innumerable galaxies throughout the entire universe, the Lord continually thinks about each of our lives and needs. It's astonishing that the God of the universe, with innumerable worlds to care for, has the *interest,* much less the divine *capacity,* to watch our individual lives and think about "the thoughts and intents of our hearts" (Alma 18:32; see also D&C 33:1; 88:109). But as Ammon said, "Now my brethren, we see that *God is mindful* of every people, whatsoever land they may be in" (Alma 26:37). He is mindful of us because he's always thinking about us and pondering how he can best help us make the journey back to him. "Many, O Lord my God, are thy wonderful works which thou hast done, and *thy thoughts which are to us-ward* . . . cannot be reckoned up . . . if I would declare and speak of them, *they are more than can be numbered"* (Ps. 40:5).

It isn't hard to picture him thinking about the apostles and prophets, but his thoughts and attention are directed equally to the very least of us. This is true even though, in comparison to them, "I am poor and needy; yet the Lord thinketh upon me" for *"he remem-bereth every creature* of his creating" (Ps. 40:17; Mosiah 27:30).

Speaking on the subject of God's loving attention to the details of our individual lives, Elder Paul H. Dunn said, "We must remember that our struggles here are not too trivial to interest the Lord." And then, after referring to some of the immensities of space and God's innumerable creations, he continued:

> When we think that our Lord created all that . . . and understands the immensities of space that only boggle our minds, we have difficulty believing that his interest could ever devolve on something as minute as our little pain or concerns. Well, our frustrations and disappointments may be just pinpricks in the eternal scheme of things, but since they do not seem that way to us, they do not seem that way to the Lord. The Lord is waiting to help you cope if you will lay your human sized needs at his divine feet *(Ensign,* May 1979, pp. 7-9).

> Can a woman forget her sucking child, that she should not have compassion on the son of her womb? Yea, they may forget, yet will I not forget thee, O house of Israel (1 Ne. 21:15; also Isa. 49:15).

Incredible as it seems to we who are so dependent upon sleep and must live our lives in cycles, the Lord's attention upon us is unceasing, for "He that keepeth Israel shall neither slumber nor sleep," so that "the ways of man are before the eyes of the Lord, and he pondereth all his goings" (Ps. 121:4; Prov. 5:21). As Thomas S. Monson emphasized, "Brethren, the Lord knows each of us. Do you think for a moment that He who notes the sparrow's fall would not be mindful of our needs and our service? We simply cannot afford to attribute to the Son of God the same frailties which we find in ourselves" *(Ensign,* Nov. 1989, p. 46).

As unworthy and undeserving as we are in our fallen condition, if God directs so much of his mind's attention to us, surely we can respond to this love by keeping our thoughts clean and devoting time to remembering and appreciating him. "And now, O man, remember, and perish not" (Mosiah 4:30).

IMPORTANT NOTES ON BRAIN CHEMISTRY
For every thought and feeling in the *spirit mind,* there exists a neurochemical equivalent in the *physical brain.* The way we think, the way we feel, and the way we act all happen because of chemical actions in the physical brain, which occur in *response* to the thoughts and

instructions coming from our spirit minds. These complex chemical reactions are dependent upon a delicate balance of millions of chemical molecules (called neurotransmitters), all working together with specialized brain cells, to communicate and act on the thoughts, feelings, and instructions coming from the spirit mind.

We learned earlier that the physical brain consumes about 20% of our body's nutrients. Recent discoveries by brain scientists have shown that a lack of proper nutrients to the brain (essential amino acids that fuel the production of crucial neurotransmitters) can result in the chemical imbalances that influence mood and attention disorders like depression, stress, anxiety, irritability and hostility, sleeping difficulties, and ADD/ADHD.

We now know that if the physical brain chemistry is not healthy and properly balanced, we cannot experience thoughts and feelings the way God intended us to. Our lives will be confused, and the problems, weaknesses, and temptations we're struggling with will seem overpowering. This is one reason that tens of millions of desperate people are being medicated with brain chemistry prescriptions—as I was for seventeen years.

I discovered that once the chemical balance is restored through proper *nutritional supplementation,* most people are once again able to live the kind of life they were *wanting* to but *couldn't* because of the inability of their physical brains to carry out the desires and instructions of the spiritual mind.

If you know anyone dealing with such issues—anyone who might benefit from the information that I have found so helpful—please feel free to contact me by e-mail at **StevenCramer@juno.com,** or through the publisher by mail.

Chapter Ten

THE HEART

Imagine going to a hardware store and asking for a pump about the size of your fist. This device must be able to pump thick fluid through itself at a speed of over two miles per hour and have sufficient power to circulate that flow of liquid through 6,000 miles of tubing in less than one minute. And it must do this unceasingly, twenty-four hours a day, for the next seventy to eighty years, without ever shutting down for maintenance.

Automatically, whether we think about it or not, the heart we often take for granted does all this and much more. More than seventy times a minute, from before we're born until the moment we die, our heart muscles squeeze hard and force blood to surge under pressure into the arteries, to be dispersed throughout the body to feed and nourish over three trillion cells.

The Lord has instructed us to "keep thy heart with all diligence; *for out of it are the issues of life*" (Prov. 4:23). This is true both physically as well as spiritually, for just as the heart controls the body's physical health, it also works with the mind to control a person's spiritual life through emotions, values, and priorities.

The reason "*a man's heart deviseth his way*" spiritually (Prov. 16:9) is that we usually act (or react) by *feelings* rather than premeditated *thought*. This means that we do what we *feel* like doing (at that particular moment of choice) instead of what we might do if we took time to think and determine what we *should* do. This is not as it should be.

But it's the mortal reality of imperfect beings that whenever we sin, it's either because of wrong *thinking* or wrong *feeling*. As Elder Neal A. Maxwell stated, "what we insistently desire, over time, is what we will eventually become and what we will receive in eternity" (*Ensign*, Nov. 1996, p. 21).

If we're to rise above our fallen nature and attain the Savior's image, the victory must be gained both in our thoughts and emotions, because "as he thinketh in *his heart*, so is he" (Prov. 23:7). This important scripture links the relationship between our mental thoughts, or what we know and believe we should do, to our actual behavior, which is, in fallen man, almost always a combination of what we think in our minds and what we feel in our hearts. Of course, we don't literally *think* in our physical heart-pump, but this concept is often expressed in scriptures because the heart is the spiritual symbol that represents the *combination* of thoughts mixed with our feelings, desires, and intents.

BRAIN FUNCTIONS IN THE HEART

Many functions we normally associate with the brain and with mental process are paralleled in the heart. Hundreds of scriptures refer to thinking, speaking, pondering, and praying in our hearts. For example: "Now Haman *thought in his heart* . . ." (Esth. 6:6). "Behold, they *say* and *think* in their hearts . . ." (D&C 10:16). "Jesus knowing their thoughts said, Wherefore *think ye evil in your hearts?*" (Matt. 9:4). "Repent therefore of this thy wickedness, and pray God, if perhaps *the thought of thine heart* may be forgiven thee" (Acts 8:22). "Beware that there be not a thought in thy wicked heart" (Deut. 15:9).

The scriptures not only refer to "thinking" in our hearts, but also "speaking" in them. For example: "Esau *said* in his heart. . . ." (Gen. 27:41) and, "Now, after the lord had withdrawn from speaking to me, and withdrawn his face from me, *I said in my heart* . . ." (Abr. 2:12). "Now Hannah, *she spake in her heart*; only her lips moved, but her voice was not heard" (1 Sam. 1:13).

Perhaps *speaking* in our hearts is not literally "talking" so much as *reasoning* as we weigh values and consider alternatives. This is illustrated by Nephi's dilemma and reasoning when he was instructed to kill and went through an agonizing thought process that involved deep and conflicting feelings that had to be resolved before he could obey: "And it came to pass that I was constrained by the Spirit that I should kill Laban; *but I said in my heart*: Never at anytime have I shed the blood of man. And I shrunk and would that I might not slay him" (1 Ne. 4:10).

Similarly, when the wicked want to justify their acts of disobedience, they reason and rationalize in their hearts that it's their right to do as they please, and they create excuses to justify violating their conscience. "*Then they say in their hearts*: this is not the work of the Lord, for his promises are not fulfilled" (D&C 58:33). "Wherefore [how] doth the wicked contemn [ignore and reject] God? he hath *said in his heart*, Thou wilt not require it" (Ps. 10:13). "He hath said in his heart, God hath forgotten: he hideth his face; he will never see it" (Ps. 10:11).

The scriptures often refer to this heartfelt process of reasoning, weighing, and self-talk as "pondering." Academically, this word would be associated with a *mental* process, but in the scriptures it almost always refers to the heart rather than the mind, because the Lord wants us to reflect on our feelings and values as well as our thoughts, as did Mary, who "kept all these things, and pondered them in her heart" (Luke 2:19). "Commune with your own heart" (Ps. 4:4).

Deep contemplation in the heart has been the catalyst for some very significant revelations. For example, Nephi's magnificent vision of the Tree of Life, the birth and ministry of the Savior in Jerusalem, and much more, all came to him "*as I sat pondering in mine heart.*" (See 1 Ne. 11-14.) The precious revelations regarding the spirit world were given to President Joseph F. Smith as he "sat in [his] room pondering over the scriptures; and reflecting upon the atoning sacrifice that was made by the Son of God, for the redemption of the world" (D&C 138:2-3).

While one writer reports "*I communed with mine own heart*" (Eccl. 1:16), Isaiah laments that none of his people even "considereth in his heart" (Isa. 44:19). Each Sabbath, as we renew our covenants by partaking of the sacrament, we're asked to pledge to make an effort to remember Christ in our daily lives. Perhaps the purpose of this mental and emotional memory covenant is best symbolized by Nephi's words: "Behold, my soul delighteth in the things of the Lord; and *my heart pondereth continually* upon the things I have seen and heard" (2 Ne. 4:16).

Not only are we commanded to "commune" with our hearts and use them to "ponder" spiritual things, but we're also commanded to pray in (and with) our hearts. Mormon counseled, "Wherefore, my beloved brethren, pray unto the Father *with* all the energy of heart" (Moro. 7:48). And Jesus commanded the righteous Nephites "that they should not cease to pray *in* their hearts" (3 Ne. 20:1). He has repeated the same command to us: "And again, I command thee that thou shalt pray vocally *as well as in thy heart*" (D&C 19:28). "Trust in him at all times; ye people, *pour out your heart before him*" (Ps. 62:8).

HEALTH FACTORS

The range of concerns the Lord has revealed concerning our hearts is so vast that it would be more deserving of an entire book than a mere chapter. The following chart presents some of the Lord's concerns about the emotions, values, and priorities we hold within our hearts. (Please note that the references given are only samplings and are not intended to be a complete reference of quotations.)

Characteristics

Having a "largeness of heart" (1 Kgs. 4:29)

Having "a thankful heart" (D&C 62:7)

"With a sincere heart" (Moro. 10:4)

Having "a willing heart" (Ex. 35:5)

"With an honest heart" (D&C 8:1)

Having "lowliness of heart" (1 Ne. 2:19)

"The contrite heart" (D&C 21:9)

"In solemnity of heart" (D&C 100:7)

Having a "tender heart" (2 Kgs. 22:19)

"Let thine heart be softened" (D&C 121:4)

Having "an understanding heart" (1 Kgs. 3:9)

Having a "perceptive" heart (Deut. 29:4)

Having "a trembling heart" (Deut. 28:65)

Having "a fearful heart" (Isa. 35:4)

Values

"He had set his heart upon . . ." (Ether 8:7)

"The wish of mine heart" (Alma 29:1)

"Whatsoever your heart desireth" (Hel. 13:27)

"Treasure up in your heart" (D&C 11:26)

"The desires of his own heart" (Mosiah 11:2)

"There will your heart be also" (3 Ne. 13:21)

"My heart delighteth in righteousness" (2 Ne. 9:49)

Priorities

"The thoughts of my heart" (Alma 18:20)

The "intents of his heart" (Alma 12:7)

"The affections of thy heart" (Alma 37:36)

"With real intent of heart" (Moro. 7:9)

"He laid the plan in his heart" (Alma 47:4)

"My heart doth magnify his holy name" (2 Ne. 25:13)

Emotions

"The feelings of his heart" (2 Ne. 4:12)

"Lift up thy heart" (D&C 25:13)

"Let not your heart be troubled" (John 14:27)

"Let your heart be comforted" (D&C 101:15)

"Let your heart be glad" (D&C 79:4)

Having "joy of heart" (Prov. 65:14)

Having "a merry heart" (Prov. 17:22)

Having "sorrow of heart" (Isa. 65:14)

Being troubled by a "heavy heart" (Prov. 25:20)

Healthy Hearts

Having "a sound heart" (Prov. 14:30)

Having "a true heart" (Heb. 10:22)

Having "integrity of heart" (Gen. 20:5-6)

Having "a clean heart" (Ps. 51:10)

Being "pure in heart" (Matt. 5:8)

Being "upright in heart" (D&C 61:16)

"With holiness of heart" (Mosiah 18:12)

Attaining "a perfect heart" (2 Kgs. 20:3)

Unhealthy Hearts

Those who "err in their heart" (Ps. 95:10)

"Thy heart is not right in the sight of God" (Acts 8:21)

Allowing "the devil to lead away your heart" (Alma 39:11)

Having a "foolish heart" (Rom. 1:21)

"Without stubbornness of heart" (Alma 32:16)

Having a "stony heart" (Ezek. 11:19)

"The blindness of their heart" (Eph. 4:18)

Having a deceitful heart (Prov. 12:20)

"Hypocrites in heart" (Job 36:13)

"The foolish imaginations of his heart" (1 Ne. 2:11)

"In the pride of his heart" (Alma 1:6)

"He exalted himself in his heart" (D&C 63:55)

Having a "double heart" (1 Chr. 12:33)

A heart that faints (Gen. 45:26)

Responsive Hearts

"That a seed may be planted in your heart" (Alma 32:28)

"They were pricked in their heart" (Acts 2:37)

"His heart was moved" (2 Ne. 17:2)

"Sunk deep into my heart" (Enos 1:3)

"Have full sway in your heart" (Alma 42:30)

"With an open heart" (D&C 64:22)

"Lay it to heart" (D&C 58:5)

"They were cut to the heart" (Acts 5:33)

Having "a broken heart" (Ps. 34:18)

"A change of heart" (Alma 5:26)
"A mighty change wrought in his heart" (Alma 5:12)

Committed Hearts

Having an "established" heart (Ps. 112:8)

"With all thy heart" (D&C 59:5)

"With full purpose of heart" (3 Ne. 10:6)

Being "stedfast in his heart" (1 Cor. 7:37)

"With singleness of heart" (D&C 36:7)

"They were of one heart" (Moses 7:18)

"With all the energy of heart" (Moro. 7:48)

THE FALLEN HEART

"But this people hath a revolting and a rebellious heart" (Jer. 5:23).

It's not only our *outward* sins of behavior that trouble the Lord, but also our inward sins of improper *feelings*. When we inherited the fall and the natural-man dilemma I described in Chapter Eight, it wasn't just our physical *bodies* that placed us in enemy status against God and godliness, but also our *minds* and *hearts*, or in other words, our thoughts and feelings, that "began from that time forth to be carnal, sensual, and devilish" (Moses 5:13). "And God saw that the wickedness of man was great in the earth, and that every imagination of the thoughts of his heart was only evil *continually*" (Gen. 6:5).

About 32,000 babies born in America each year enter mortality with some form of *physical* heart defect, but virtually *every person* entering mortality has *spiritual* heart defects. The dangers that lie within the emotions of our powerful, fallen, natural-man hearts are a reality that must be faced and conquered by *every* descendent of Adam. "And the Lord spake unto Adam, saying: Inasmuch as thy children are conceived in sin, even so when they begin to grow up, *sin conceiveth in their hearts*" (Moses 6:55).

This scripture doesn't mean that the act of conception within the bounds of marriage is sinful. It means, rather, that all children are born into the environment of a sinful world, which leads them *naturally* into sinful ways unless righteous parents intervene and teach them a higher way. Because mankind is fallen, "there are many devices in a man's heart," along with many "evils and designs" that must be overcome if our hearts are to become pure, Christlike, and trustworthy. (See Prov. 19:21; D&C 89:4.) Because man is fallen, it's customary and natural to "set their heart [desires] on their iniquity" (Hosea 4:8). Indeed, scripture warns, "He that trusteth in his own heart is a fool" (Prov. 28:26) because "the [unredeemed] heart is *deceitful above all things, and desperately wicked*" (Jer. 17:9).

Most victims of heart disease encounter drastic changes in their health, energy, and ability to continue life as they once knew it. Their entire health focus must center on living within the constraints imposed by the limitations of the defective heart. And so it is spiritually. Perhaps the greatest of all threats to our eternal happiness are the limitations imposed by the compelling desires and passions of our own fallen, natural-man hearts. "For out of the heart proceed evil thoughts, murders, adulteries, fornications, thefts, false witness, blasphemies: These are the things which defile a man" (Matt. 15:19-20).

SPIRITUAL "HEART ATTACKS"

It takes several hours to get just a couple of gallons of drinking water filtered through my expensive, hi-tech reverse osmosis filter system, yet my fist-sized heart can pump over *twenty gallons* of blood through my liver every hour to remove the waste materials and impurities from my blood. During normal activity, the heart beats about 4,200 times per hour. That's more than 100,000 times a day; about three million times a month, or two and a half billion beats in a lifetime. (No man-made material can be flexed so frequently without breaking.) During each of these normal beats, your heart pumps about two and a half ounces of life-sustaining blood. That's about six quarts a minute, or two thousand gallons a day—which could fill four or five average-sized hot tubs or spas. (During strenuous exercise, the heart can pump as much as thirty quarts a minute!)

The heart must pump this enormous flow of blood through some 6,000 miles of blood vessels, most of which are tiny capillaries with a diameter about thirty times smaller than a human hair. The heart pumps with such power that it may take less than a minute for your blood to circulate around your entire body. And yet, with no thought on our part, more than 1,000 times a day the heart sends each blood cell on this journey so that every cell in the body can receive nourishment and oxygen and have its waste materials carried away.

If this blood flow is interrupted to any part of the body, even for just a few minutes, that region will be seriously damaged and may die. It's estimated that more than 57 million Americans have some form of cardiovascular disease. These heart diseases kill over one million people a year, about 40% of all deaths in America. (By comparison, accidental deaths only kill about one-tenth this many.) In fact, heart attacks are the single leading cause of death. In America there are more than 4,000 heart attacks a day. About one-third of these attacks result in death; that's more than 600,000 people a year.

The Savior warned us that one of the major "signs" of the last days would be that "all things shall be in commotion; and [that] *men's hearts shall fail them*" (D&C 88:91). I believe this prophecy pertains not only to *physical* heart failure but also to *spiritual* heart attacks, which are far more common than the physical ones. (See also D&C 45:26; Moses 7:65-66.)

One of the biggest dangers of heart disease is that most people aren't aware of it until it has reached a critical stage. This is because physical heart diseases involve a long-term, progressive, and cumulative process that quite often produces no discernable symptoms until the disease is quite advanced. For example, as many as 50 million Americans have high blood pressure, yet more than one-third of them are completely unaware of it. About 40,000 will die each year because of this lack of awareness.

Every day people are dying *spiritually* because they're unaware of their spiritual heart diseases. Just as our *physical* lives depend on the heart's unceasing and automatic contractions, our *spiritual* lives depend on the heart's unceasing commitment to righteous desires and priorities.

One common reason for death by heart failure is the denial that anything is wrong. The resulting delay of examination and treatment, even after a person becomes aware of the symptoms and warning signs, can prove fatal. For example, when a person experiences mild to severely crushing chest pains from a first heart attack, the symptoms often pass in a few moments. Rather than rush in for evaluation and treatment, many people ignore the seriousness of the warning signs and go on about life, hoping they can get away with it. (They can't!)

And so it is with spiritual heart failure. We find ourselves actually considering breaking or compromising our vows. We become apathetic about our commitments and promises. These are alarming danger signs, yet we ignore them. Just as a person with a physically damaged heart loses energy and concern for life's pressures and duties, losing spiritual energy and concern should be a warning signal that things are not right with our hearts spiritually. When performance of our family and gospel duties becomes more painful to us than neglect, we're in serious danger of heart failure and possible spiritual death and should undertake an immediate evaluation of our values and priorities.

Many problems can develop in our physical heart and circulatory system. Each defect produces specific symptoms, such as shortness of breath, chest, shoulder, and arm pains, poor circulation, lack of energy, erratic heart rate, and so forth. Doctors try to detect pending heart diseases by watching for these warning signs; they listen to our heartbeat and they test our blood pressure and cholesterol levels.

We can recognize physical heart attacks by the sudden onset of crushing pains in the chest, which often radiate to the shoulder, arms, and jaw. Spiritual heart attacks (in which "men's hearts shall fail them" D&C 88:91) can be recognized by symptoms like an urge to break our covenants, to make evil choices, or to hurt others as we strike out in anger, revenge, or an unwillingness to forgive.

The following chart presents some other symptoms of spiritual heart disease that we can watch for so that we don't suffer spiritual heart attacks by becoming "a people that do err in their heart[s]" and whose "heart[s are] not right in the sight of God" (Ps. 95:10; Acts 8:21).

Having an "evil heart" (Jer. 7:24)	Having "a rebellious heart" (Jer. 5:23)
Having "a wicked heart" (Prov. 25:20)	Being "uncircumcised in heart" (Acts 7:51)
Having a "perverse heart" (Prov. 12:8)	

Having an "obstinate" heart (Deut. 2:30)

A "naughtiness" of heart (1 Samuel 17:28)

Having "a despiteful heart" (Ezekiel 25:6)

Having an "impenitent heart" (Rom. 2:5)

Being a "backslider in heart" (Prov. 14:14)

A heart that turns from God (1 Kgs. 11:2-9)

"Their heart is far from me" (Mark 7:6)

Having "a double heart" (Ps. 12:2)

"Their heart is divided" (Hosea 10:2)

"The haughtiness of his heart" (Jer. 48:29)

Being "proud in heart" (Prov. 16:5)

Having "blindness of heart" (D&C 58:15)

Having a "doubtful heart" (D&C 58:29)

Having a "bitterness of heart" (Ezek. 27:31)

"The murmurings of his heart" (D&C 75:7)

Having a "despising" heart (2 Sam. 6:1)

SATANIC "HEART ATTACKS"

"Satan seeketh to turn their hearts away from the truth, that they become blinded and understand not the things which are prepared for them" (D&C 78:10).

Spiritual heart attacks come not only from our natural, fallen nature, but also from Satan, whose goal is to oppose the Lord's work of redemption by deceiving and leading us astray so that we become miserable as he is. (See 2 Ne. 2:18-27.) There are many fronts to the devil's war against Christ, but one of his greatest efforts is to capture *the feelings of our hearts*. Why would this be? Because he knows that "a man's heart deviseth his way" (Prov. 16:9). He knows that polluting the feelings and values we hold in our hearts will have a far greater chance of destroying our spiritual lives than merely confusing our minds. That's why the scriptures say that he will seek to "*rage in the hearts* of the children of men, and stir them up to anger against that which is good" (2 Ne. 28:20). "And thus Satan did *lead away* the hearts of the people to do all manner of iniquity" (3 Ne. 6:16).

With thousands of years to perfect his strategies, Satan has many "cunning plans which he hath devised to *ensnare the hearts* of men" (Alma 28:13). To "ensnare" means to capture or catch, as if in a snare.

Thus, through various cunning enticements, he captures our emotions and our will, "*leading away* the hearts of the people, tempting them and causing them that they should do great wickedness in the land" (3 Ne. 2:3). Two terms used by scripture to describe Satan's influence on our hearts are having a great *hold* on them, and having *power over* them.

SATAN'S HOLD ON OUR HEARTS

"But Amulek stretched forth his hand, and cried the mightier unto them, saying: O ye wicked and perverse generation, why hath Satan got *such great hold* upon your hearts? Why will ye yield yourselves unto him that he may have power over you?" (Alma 10:25). In this scripture, Amulek revealed that the danger of allowing Satan to have a "hold" on our hearts is the control he gains over our emotions, and therefore over our choices. Scriptures almost always refer to his influence over our desires and emotions as a "great hold." As Mormon reported, Satan "had got *great hold* upon the hearts of the Nephites; yea, insomuch that they had become exceedingly wicked" (Hel. 6:31; see also 16:23; Alma 8:9). Or, as the Lord reported in our own time: "Satan has *great hold* upon their hearts; he stirreth them up to iniquity against that which is good" (D&C 10:20).

SATAN'S POWER OVER OUR HEARTS

Once we allow Satan to get a "hold" on the values and emotions we have in our hearts, we've "given ourselves away" to him, giving him "power over" our hearts. As Nephi lamented to his wicked people:

> And because of my mourning and lamentation ye have gathered your-selves together, and do marvel; yea, ye ought to marvel because *ye are given away* that the devil has got so great hold upon your hearts. Yea, *how could you have given way* to the enticing of him who is seeking to hurl away your souls down to everlasting misery and endless wo? (Hel. 7:15-16).

To lose control over our heart's emotions is to lose control over our agency. When we deliberately, persistently, over a period of time choose the ways of Satan over the ways of the Lord, allowing him to obtain that great hold on the emotions and values in our hearts, we eventual-ly reach a point of such enslavement to his influence that we become as Babylon, "in whose hearts the enemy, *even Satan, sitteth to reign*" (D&C 86:3). Such was the cause and case that resulted in the destruction of the Jaredites: "And behold, the Spirit of the Lord had ceased striving with them, and Satan had *full power over* the hearts of the people; for they were *given up* unto the hardness of their hearts, and the blindness of their minds that they might be destroyed" (Ether 15:19).

Each time we give in to his evil enticements, we open the door to his increasing influence. For example, those who allow their thoughts to wander into sexual fantasies find their thoughts and emotions more and more difficult to control. Having deliberately opened the door to unworthy thoughts, the person has unknowingly invited the unclean devils to enter and take control of their thoughts and emotions. These evil spirits will eagerly guide our thoughts and feelings into progressively downward spirals of filthiness until they compel us to seek wicked fulfillment by acting out the lewd desires.

We can also give Satan power over our hearts by participation in the use of mind-altering drugs, alcohol, pornography, masturbation, petting, and other lewd acts of sexual lust. These unfortunate choices open the door to control over the desires and passions of our hearts as surely as if we'd posted a welcome sign. As Elder Melvin J. Ballard said, "Secret weaknesses and vices leave an open door for the enemy of your souls to enter, and he may come in and take possession of you, and you will be his slave" (*The New Era*, March 1984, p. 38). As we struggle against the enslaving possessions of evil habits of thought and emotion, we try to resist, but fail, falling back into endless cycles of defeat, repeating the sins we hate. We're only seeing the determination of our enemies to hold us in captivity to our fallen minds and hearts. So it was in the Book of Mormon: "The people having been delivered up for the space of a long time to be carried about by the temptations of the devil, whithersoever he desired to carry them, and *to do whatsoever iniquity he desired they should*" (3 Ne. 6:17).

Elder Richard G. Scott said, "It is as though Satan ties strings to the mind and body so that he can manipulate one like a puppet" (*Ensign*, May 1986, pp. 10-11). How important, then, to "pray always lest that wicked one *have power in you*, and remove you out of your place" (D&C 93:49). The Lord warned, "Remember that he that persists in his own carnal nature, and goes on in the ways of sin and rebellion against God, remaineth in his fallen state and *the devil hath all power over him*" (Mosiah 16:5).

Satan and his spirits have great powers of persuasion and, if we will allow it, even possession. But there are boundaries beyond which they are not allowed to go—unless we invite them—and that includes power over our hearts. As Joseph Smith said, "The devil has no power over us only as we permit him" (*Teachings of the Prophet Joseph Smith*, sel. Joseph Fielding Smith [Salt Lake City: Deseret Book Co., 1938], p. 181). Thus, "God is faithful, who will not suffer you to be tempted above that [which] ye are able [to withstand], but will with the temptation also make a way to escape, that ye may be able to bear it" (1 Cor. 10:13).

> Satan and his aides no doubt may know of our inclinations, our carnal tastes and desires, but *they cannot compel a righteous person to do evil* if he seeks help from the Lord. Too many try to blame Satan when in reality the fault lies within themselves because they yield to his enticements (ElRay L. Christiansen, *The New Era*, July 1975, p. 49).

> Rejoice, O my heart, and *give place no more* for the enemy of my soul (2 Ne. 4:28).

Nephi prophesied of a time to come when "Satan shall have power *over* the hearts of the children of men no more" (2 Ne. 30:18). We can bring that time to pass right now by yielding our hearts to God instead of Satan, for it is "because of the righteousness of his people [that] Satan has no power . . . for he hath no power *over* the hearts of the people [when] they dwell in righteousness, and the Holy One of Israel reigneth" (1 Ne. 22:26). Moroni said that he was "commanded to write these things that evil may be done away, and that the time may come that Satan may have no power *upon* the hearts of the children of men but that they may be persuaded to do good continually" (Ether 8:26). "And Satan shall be bound, that he shall have no place in the hearts of the children of men" (D&C 45:55).

SPIRITUAL ATHEROSCLEROSIS

"Yea, they made their hearts *as an adamant stone*, lest they should hear the law" (Zech. 7:12).

"They had hardened their hearts against him, that they *had become like unto a flint*" (2 Ne. 5:21).

"His heart is as firm as a stone" (Job 41:24).

Atherosclerosis is the disease commonly called "hardening of the arteries." It's caused by the gradual accumulation of cholesterol plaque inside our blood vessels over a long period of time. To compensate for our unhealthy habits, the Lord has built so many safety factors into our physical bodies that our arteries can still deliver adequate blood to maintain normal function even when they are blocked by as much as 80%!

But when the blockage reaches 90-95%, the blood pressure finally causes a tiny rupture in the plaque blockage to break off, allowing a clot and spasm to close the vessel, causing sudden and severe damage. If this blockage occurs in a vessel leading to the brain, the result is a stroke. If it occurs in a vessel leading to the heart, it causes a heart attack, resulting in death or severe heart damage. If the stroke or heart attack doesn't kill us, it will certainly leave us with pain, and with mental and physical limitations that restrict our ability to live life to its fullest potential.

There are also mental, emotional, and behavioral factors that can harden our hearts *spiritually* until they leave us with severe spiritual limitations that restrict our ability to fulfill the purposes of mortality,

Scriptures describe some very intense feelings of the heart. We're told, for example, of righteous people whose "hearts *were swollen with joy,* unto the gushing out of many tears, because of the great goodness of God" (3 Ne. 4:33), while others had become so hardened that they "were *past feeling,* that [they] *could not feel* his words" (1 Ne. 17:45; see also Eph. 4:19; Moro. 9:20). It's estimated that more than 60 million Americans are living with physical and emotional limitations caused by unsolved heart disease, but virtually 100% of humanity is struggling with the limitations imposed by spiritual heart disease, especially hard hearts. To achieve his image in our lives, our hearts must be so soft and yielding to him that his word will "have full sway in [our] heart" (Alma 42:30).

Let's review five of the sequential, progressive consequences of stubborn and hardened hearts that we bring upon ourselves when we refuse to "soften" and yield our hearts to God.

1. Loss of Spiritual Discernment

Fully understanding and successfully living gospel truths, the purpose and perplexities of life, and the ways of our Heavenly Father requires the spiritual discernment and understanding that comes from the influence of the Holy Ghost. The most immediate consequence of hardening our hearts against God is loss of that Spirit. We're told, for example, "that the Spirit of the Lord began to withdraw from the Nephites, because of the wickedness and the hardness of their hearts" (Hel. 6:35).

When the Lord withdraws his influence from us, it isn't because he no longer loves us, but because we've forced him to do so by the unwise use of our agency. "There are many that harden their hearts against the Holy Spirit that it hath no place in them; wherefore, they cast many things away which are written and esteem them as things of naught" (2 Ne. 33:2). If the Lord's Spirit and influence is to dwell in our hearts, it requires hearts that are soft and yielding. There's just no room for him in a heart that's hardened by sin or lack of caring. "But they hardened their hearts and could not endure his presence" (D&C 84:24). "But your iniquities have separated between you and your God, and [it is] your sins [which] have hid his face from you, that he will not hear" (Isa. 59:2).

When we shut the Lord out of our lives by hardening our hearts, we lose the discernment to recognize and appreciate his attempts to bless us and improve our lives. We read in amazement at the blindness of the Nephites, who "began to forget those signs and wonders which they had heard, and began to be less and less astonished at a sign or wonder from heaven, insomuch that they began to be hard in their hearts, and blind in their minds, and began to disbelieve all which they had heard and seen" (3 Ne. 2:1). Yet we, today, are just as blind and hardened. Even though the Lord has told us of the upheavals of nature that will announce the last days, we, too, are "less and less astonished" as the daily news recounts global weather events setting new records of devastation. Numbly, we shrug and take these warning signs of the end drawing near for granted.

For after your testimony cometh the testimony of earthquakes, that shall cause groanings in the midst of her, and men shall fall upon the ground and shall not be able to stand.

And also cometh the testimony of the voice of thunderings, and the voice of lightnings, and the voice of tempests, and the voice of the waves of the sea heaving themselves beyond their bounds.

And all things shall be in commotion; a*nd surely, men's hearts shall fail them;* for fear shall come upon all people (D&C 88:89-91).

If our hearts are so hardened, our eyes so blinded, and our ears so deafened that we can't see and hear the thunderous roar of these signs of the times announcing the end of time, how could we possibly expect to have the spiritual discernment to see the more subtle workings of the Lord in our personal lives? "For *they considered not* the miracle of the loaves: *for their heart was hardened"* (Mark 6:52) and "they knew not that God could do such marvelous works, for they were a hard-hearted and a stiffnecked people" (Alma 9:5). "And behold, the Spirit of the Lord had ceased striving with them, and Satan had full power over the hearts of the people; for they were given up unto the hardness of their hearts" (Ether 15:19).

2. Loss of Spiritual Understanding

When we lose the Spirit by hardening our hearts, we also close our minds and lose understanding of gospel principles as we gravitate toward the darkness. "And now, did they understand the law? I say unto you, Nay, they did not all understand the law; and this because of the hardness of their hearts" (Mosiah 13:32; see also Alma 33:20). "And they that will harden their hearts, to them is given the lesser portion of the word *until they know nothing concerning his mysteries*; and then they are taken captive by the devil, and led by his will down to destruction" (Alma 12:11).

Just as we offend the Holy Spirit by shutting out its influence with hard and unyielding, uncaring hearts, the prophets, too, are sometimes forbidden to teach us with the Lord's word. They're required to withdraw their messages and counsel until we're more receptive. Mormon stated, "I was forbidden to preach unto them, because of the hardness of their hearts" (Morm. 1:17). "And it came to pass that the people hardened their hearts, and would not hearken unto their words; and the prophets mourned and withdrew from among the people" (Ether 11:13).

The Lord has explained the relationship between the hardness or the softness of our hearts and the receiving of spiritual understanding in these two verses: "*Because of the hardness of the hearts* of the people of the Nephites, except they repent I will take away my word from them, and I will withdraw my Spirit from them, and I will suffer them no longer" (Hel. 13:8).

On the other hand: "If ye will *not harden* your hearts, and ask me in faith, believing that ye shall receive, with diligence in keeping my commandments, surely these things shall be made known unto you" (1 Ne. 15:11).

3. Losing Christ from Our Lives

It's inevitable that as we lose the Spirit and spiritual understanding of gospel principles, we lose the companionship and blessing of the Savior as well. When our hardened hearts have shut out the Lord's spirit from guiding us, we're left to rely upon ourselves instead of him. Thus, "being hard in their hearts, therefore they did not look unto the Lord as they ought" (1 Ne. 15:3). There's no way a person can receive Christ with a hard heart. There's just no room for him until we're soft and yielding, recognizing that his will is superior to our own. "They will reject him, because of their iniquities, and the hardness of their hearts, and the stiffness of their necks" (2 Ne. 25:12).

4. We Fall into the Sinful Ways of "The Natural Man"

How could we possibly live righteously when our stubborn, hardened, and resisting hearts have driven the Holy Spirit and the Savior himself from our lives and left us without divine guidance or support? Thus, "he that hardeneth his heart shall fall into mischief" (Prov. 28:14). Like a person who refuses to stop smoking or change an unhealthy diet after a life-threatening heart attack, a person with a hard heart just gives up and surrenders to natural and wicked instincts. "And behold, the Spirit of the Lord had ceased striving with them, *and Satan had full power over the hearts of the people*; for they were given up unto the hardness of their hearts" (Ether 15:19).

5. We Reap the Natural Consequences of Our Indifference

Just as a physician is grieved by the choices of people whose lifestyles result in life-threatening heart disease, the Lord is also grieved by unfortunate choices that result in spiritually hardened hearts. As New Testament writers reported, Jesus was "grieved for the hardness of their hearts" (Mark 3:5). The Savior will always be grieved, and at times even angered, by choices and situations that harm his brothers and sisters, especially when we continually harden our hearts in spite of his attempts to teach us a better way. "I am angry with this people, and my fierce anger is kindled against them; for their hearts have waxed hard" (Moses 6:27; see also Mark 16:14). "And now, my brethren, seeing we know these things, and they are true, let us repent, and harden not our hearts, that we provoke not the Lord our God to pull down his wrath upon us" (Alma 12:37).

Because the Lord loves us so much, he now allows us to reap the bitter and painful consequences we bring into our own lives by resisting and withdrawing from his influence. He allows it in the hope that these painful tutoring situations will persuade us to soften and yield our hearts to him. "Nevertheless, when they shall have received the fullness of my gospel, then if they shall harden their hearts against me *I will return their iniquities upon their own heads*, saith the Father" (3 Ne. 20:28). "But those who harden their hearts in unbelief, and reject it, it shall turn to their own condemnation" (D&C 20:15).

> And thus we can plainly discern, that after a people have been once enlightened by the Spirit of God, and have had great knowledge of things pertaining to righteousness, and then have fallen away into sin and transgression, they become more hardened, *and thus their state becomes worse than though they had never known these things* (Alma 24:30).

THE CAUSES OF SPIRITUAL ATHEROSCLEROSIS

Some "hardening of the arteries" occurs naturally as we age. It can also result from things like smoking, diabetes, high blood pressure, and obesity, but most of all it comes from poor nutrition, lack of exercise, and an excess of fats in our diet. When a doctor discovers a patient with hardening of the arteries, a thorough examination is immediately conducted of the patient's diet and health practices so that the causes may be discovered and eliminated. Since an emotionally hardened heart can lead to actual heart failure and spiritual death, it's important to identify the specific causes that lead to this dangerous and spiritually fatal condition. Let's consider seven of the most common causes of *spiritual* hardening that diminish people's mortal experience.

1. We Are Fallen

Many hearts are hardened simply because we're in a fallen state that makes it more *natural* to resist and turn away from God than to yield to him and grow through obedience.

> And thus we can behold how false, and also the unsteadiness of the hearts of the children of men. Yea, and we may see at the very time when he doth prosper his people . . . doing all things for the welfare and happiness of his people; yea, then is the time that they do harden their hearts, and do forget the Lord their God (Hel. 12:1-2).

2. Sin

Just as too much cholesterol will harden our arteries, deliberate sin hardens our hearts, as Alma reported of "the Nephites, who are now hardening their hearts in sin and iniquities" (Alma 37:10). "But exhort one another daily, while it is called To day; lest any of you be hardened through the deceitfulness of sin" (Heb. 3:13).

3. Unbelief

Just as people choose poor health practices because they believe they can abuse their bodies and get away with it, many of us harden our hearts because we don't believe what the Lord has said. It is both doubts and the "veil of unbelief which doth cause [us] to remain in [our] awful state of wickedness, and hardness of heart, and blindness of mind" (Ether 4:15).

> But there were many who were so hardened that they would not look, therefore they perished. Now the reason they would not look is because they did not believe that it would heal them. O my brethren, if ye could be healed by merely casting about your eyes that ye might be healed, would ye not behold quickly or would ye rather harden your hearts in unbelief, and be slothful, that ye would not cast about your eyes, that ye might perish? (Alma 33:21).

4. Pride

Many people who die of *physical* heart disease do so because they're too proud to humble themselves and submit to the necessary laws of health. Similarly, many of us harden our hearts because of a pride that makes us want to live by our will instead of God's. "But when his heart was lifted up . . . his mind hardened in pride" (Dan. 5:20).

5. Avoiding Responsibility

We also harden our hearts to shut out the word of the Lord, pretending that if we don't *receive* it, we won't be obligated to obey it. Amulek was such a man. He knew the truth and the ways of the Lord, but was resistant and disobedient until an angel changed his heart and prepared him to become a companion and assistant to Alma. His words are interesting. "I have seen much of his mysteries and his marvelous power," he said, "Nevertheless, I did harden my heart, for *I was called many times and I would not hear*; therefore, *I knew concerning these things yet I would not know*; therefore I went on rebelling against God, in the wickedness of my heart" (Alma 10:5-6). "Yea, they made their hearts as an adamant stone, lest they should hear the law, and the words which the Lord of hosts hath sent" (Zech. 7:12).

6. Adversity

Another reason we harden our hearts is because of improper response to the adversities and opposition we came here to experience. For example, Nephi experienced the same adversities as his brothers, Laman and Lemuel. But they chose to let the trials turn them bitter and hard, while Nephi was drawn to the Lord by the very same circumstances. In the following scripture, we see a contrast between a hardening and softening response:

> But behold, because of the exceedingly great length of the war between the Nephites and the Lamanites many had become hardened, because of the exceedingly great length of the war; and many were softened because of their afflictions insomuch that they did humble themselves before God, even in the depth of humility (Alma 62:41).

7. Satanic Influence

In addition to our self-imposed hardening choices, Satan also uses many strategies to harden men's hearts because of the destruction it brings to their lives by keeping them from the truth, "for he seeketh that all men might be miserable like unto himself" (2 Ne. 2:27). Thus "he hath blinded their eyes, and hardened their heart; that they should not see with their eyes, nor understand with their heart, and be converted, and I should heal them" (John 12:40). To influence our hearts he uses rumors, contentions, lies, and temptations. (See Hel. 16:22; 3 Ne. 1:22.) "And the mists of darkness are the temptations of the devil, which blindeth the eyes, and hardeneth the hearts of the children of men, and leadeth them away into broad roads, that they perish and are lost" (1 Ne. 12:17).

THE BLESSINGS OF A SOFT HEART

"Harden not your hearts" (Heb. 3:8; D&C 45:6).

While we've discussed many causes of our hearts being hardened in sin and rebellion, there seems to be only one power that can undo that disease and soften our hearts. That is the influence of the Lord. Almost every scriptural reference to this process includes his influence: for example, "The Lord did soften the heart of Ishmael, and also his household" (1 Ne. 7:5). Young Nephi, in yet another example, had to choose between the scornful indifference of his older brothers and the visionary teachings of his father. He said "I did cry unto the Lord; and behold *he did visit me, and did soften my heart* that I did believe all the words which had been spoken by my father" (1 Ne. 2:16). The converted Lamanite king, Lamoni, said, "I thank my great God that he has given us a portion of his Spirit *to soften our hearts*" (Alma 24:8).

Of course, God can't and won't change the disposition of our hearts without our permission. But when we use our agency to turn and yield our hearts to him, then his divine intervention and treatment of the heart makes the softening process possible. Job understood this when he said, "For God maketh my heart soft" (Job 23:16).

We're promised that "as many as will not harden their hearts shall be saved in the kingdom of God" (Jacob 6:4). This promise is not just for the end-of-the-world kind of salvation, but for now, today, when we need help in being all God wants us to be. For "if ye will repent and harden not your hearts, *immediately* shall the great plan of redemption be brought about unto you" (Alma 34:31). Those with soft and yielding hearts are also promised the enviable blessing of mercy as they struggle through the mistakes of mortality. "Therefore, whoever repenteth, and hardeneth not his heart, he shall have claim on mercy through mine Only Begotten Son, unto a remission of his sins" (Alma 12:33-34).

SPIRITUAL "PACEMAKERS"

"I will put my law in their inward parts, and write it in their hearts" (Jer. 31:33).

The spiritual energy needed to regulate an energetic, healthy, Christlike life largely depends on the values, priorities, and emotional commitments stored within our hearts. Every month more than 10,000 hi-tech electronic devices called "pacemakers" are surgically implanted and connected to defective hearts to regulate their beating rhythms, enabling the recipients to have more normal health and energy. The pacemaker's computer monitors the heart rate and *automatically* sends electrical signals to the heart to adjust the contractions whenever they vary beyond the programmed variables. Just as the rate at which our heart beat regulates our *physical* energy, the emotions and values implanted within our hearts regulate the *spiritual* energy and desires we have to act righteously.

In the last chapter, we discussed the importance of reading the scriptures daily to polarize our mind-set by keeping gospel truths focused in our minds. In his desire to help us treasure and commit to the proper spiritual priorities, God is attempting to lead us past mere mental awareness or comprehension of his truths and actually "implant" or internalize his word within our hearts—to function like spiritual pacemakers. This is what happened to a special man named Enos, who described his experience as he was hunting beasts and pondering his father's gospel teachings, which, apparently, he had more or less taken for granted. However, on this occasion, as he gave serious thought to them, they "sunk deep into my heart" (Enos 1:3). Because the truths finally got past the *intellectual* center in Enos's mind and took root in his heart, where powerful emotions can motivate us to action, Enos had some very compelling spiritual experiences.

Joseph Smith's heart had a similar experience. He said that when he read James 1:5, the message "seemed to enter with great force into every feeling of my heart" (JS-H 1:12). Thus the Lord pleads with us, "Let not mercy and truth forsake thee [but] *write them upon the table of thine heart*" (Prov. 3:3). "Thus knowing gospel truths and doctrines is profoundly important, but we must also come to love them. When we love them, they will move us and help our desires and outward works to become more holy" (Neal A. Maxwell, *Ensign*, Nov. 1966, p. 22).

We know that God's teachings have a more powerful impact on our lives and testimonies when conveyed to us by the power of the Spirit rather than intellectually. The reason for this is not the Spirit's power to penetrate our *minds* so much as our *hearts*, "for when a man speaketh by the power of the Holy Ghost the power of the Holy Ghost *carrieth it unto the hearts* of the children of men" (2 Ne. 33:1). This process of internalization of righteous desires and priorities within our hearts is not done *intellectually*, but by the Holy Spirit, for they are "written not with ink, *but with the Spirit of the living God*; not in tables of stone, but in fleshy tables of the heart" (2 Cor. 3:3).

We've all heard this excuse for failure: "His heart just wasn't in it." As we learned earlier in this chapter, whether we act righteously or wickedly, most of the time we do what we *feel* like doing in that moment of time, and those preferences are based *more* on what is internalized (or "written") in our hearts than what is stored in our minds. "Thy word have I hid in mine heart," said one prophet, "that I might not sin against thee" (Ps. 119:11) and because "the law of God is in his heart; *none of his steps shall slide*" (Ps. 37:31). Having God's words "written" in our hearts (internalized like a spiritual pacemaker by daily reading and pondering) provides great spiritual power because it regulates right choices *automatically*, rather than our having to struggle between the mental and emotional pull of right and wrong.

Physical fitness builders exercise often to increase their cardiovascular capacity. They know that "a sound heart is the life of the flesh" (Prov. 14:30). With satisfaction they measure each improvement in the ability of their hearts and bodies to run longer or faster, or to increase other physical endurance measurements. "Can thine heart endure, or can thine hands be strong?" (Ezek. 22:14). As we face the ever-increasing challenges of the last-day temptations, we should be nourishing our hearts and spirits daily so that our spiritual endurance increases. "Only take heed to thyself, and keep thy soul diligently, lest thou forget the things which thine eyes have seen, and *lest they depart from thy heart* all the days of thy life" (Deut. 4:9).

Just as doctors implant electronic pacemakers to regulate the function of defective hearts, Satan is trying to regulate our behavior by "whispering" lies, deceptions, discouragements, and temptations in our ears to implant unworthy feelings, desires, and values in our hearts. If we're unaware of this diabolical effort and fail to protect our minds and hearts against this dangerous influence, these diabolical suggestions can find their way into our minds and hearts, where we mistake them for our own. (See 2 Ne. 28:20-24; Hel. 13:4-5.) It's both frightening and offensive to admit that Satan can have this influence without our knowledge or consent, but he does. That's why it's so important to internalize the scriptures daily and keep ourselves in tune with the Holy Spirit.

Let's consider a few events that occurred because of Satan's power to "whisper" messages that affect emotions, values, and choices. The lies and secret combinations that Gadianton used to manipulate society "were *put into the heart* of Gadianton by that same being who did entice our first parents to partake of the forbidden fruit" (Hel. 6:26, 27-29). The crazy idea of building the Tower of Babel to get to heaven without obeying the commandments was "*put into* the hearts of the people" by Satan, who has always tried to persuade men to rely upon substitute plans, like their own skills and works, instead of the Lord and his gospel. He constantly puts questions in our hearts about why we have to seek salvation in God's way instead of other ways. (See Hel. 6:28; Gen. 11.)

We see another example in the deception Ananias and his wife practiced when Peter said, "Ananias, why hath Satan filled thine heart to lie to the Holy Ghost, and to keep back part of the price of the land?" (Acts 5:3). Satan also puts it into our hearts to hold back money, time, and service when we have opportunities to serve and build the kingdom, but end up choosing more selfish ways to use our resources.

One of the most famous things Satan ever put into the heart of a person is seen in Judas Iscariot's betrayal of Christ. Some have thought this was simply the result of his own dissatisfaction or confusion, but John testified that it was "the devil [who] put into the heart of Judas

Iscariot, Simon's son, to betray him" (John 13:2). Satan also puts suggestions into our hearts to betray our wives, families, church, and God by breaking our covenants and choosing the ways of the world over obedience, service, and sacrifice.

> We should be on guard always to resist Satan's advances. He has power to place thoughts in our minds and to whisper to us in unspoken impressions to entice us to satisfy our appetites or desires and in various other ways he plays upon our weaknesses and desires *(Answers to Gospel Questions,* comp. Joseph Fielding Smith, Jr. [Salt Lake City: Deseret Book Co., 1957-66], vol. 3:81).

Not satisfied with mere whispered messages, Satan's evil spirits also seek to penetrate the sanctuary of our hearts with their actual presence and influence. "And he shall cast out devils, or the evil spirits *which dwell in the hearts* of the children of men" (Mosiah 3:6). When we purify our hearts and fill them with the Lord's spirit so that Satan's demons can find no entry, their influence will be nullified and "Satan shall be bound, that *he shall have no place in the hearts* of the children of men" (D&C 45:55). Nephi was protective of his heart, as we should be: "And why should I yield to sin, because of my flesh? Yea, why should I give way to temptations, that the evil one *have place in my heart* to destroy my peace, and afflict my soul?" (2 Ne. 4:27).

If someone gave you a computer, you could explore its contents by examining what was stored in its memory. Apparently, through spiritual discernment, the Lord and his prophets can scan the contents of our hearts. When Abinadi testified to the priests of wicked King Noah, he said, "I read unto you the remainder of the commandments of God, *for I perceive that they are not written in your hearts*" (Mosiah 13:11). Perhaps the judgment will involve a similar search of our hearts, for we're told that our judgment will not go well if God's word is not found implanted there, stored as values and priorities. For "if we have hardened our hearts against the word, *insomuch that it has not been found in us,* then will our state be awful, for then we shall be condemned" (Alma 12:13). Thus Moses wisely instructed, "*Set your hearts* unto all the words which I testify among you this day" (Deut. 32:46) and "*lay up these my words in your heart* and in your soul" (Deut.

11:18). "And these words, which I command thee this day, shall be in thine heart" (Deut. 6:6).

SAFEGUARDING OUR HEART'S EMOTIONS

"Wherefore, let your heart be glad" (D&C 79:4; see also 29:5; 35:26).

Since one purpose of our creation is to experience joy (see 2 Ne. 2:25), learning to control the emotions felt in our hearts is important to our spiritual success and to becoming like Christ, because "a merry heart maketh a cheerful countenance" (Prov. 15:13). When the Lord instructs us to "lift up your heart and rejoice" (D&C 31:3; see also 42:69; 27:15) and to go about our daily affairs "with cheerful hearts and countenances" (D&C 59:15), it's because "by sorrow of the heart the spirit is broken" (Prov. 15:13), while "a merry heart doeth good like a medicine" (Prov. 17:22).

When the Lord counsels us to "*let* your hearts rejoice, and be exceedingly glad" (D&C 128:22), it's because he knows that cheerfulness doesn't come naturally to fallen man. Because we live in an imperfect world, there will be times of sorrow for all of us when, in spite of the commandment to "cheer up your hearts" (2 Ne. 10:23) and to "let thy heart be of good cheer before my face" (D&C 112:4), our pain will seem overwhelming. At times like these, it's easier and more natural to let the opposition and injustices of this mortal life sour our attitudes and emotions. This susceptibility to negative thinking and feeling is not something to be ashamed of; it's just a part of the natural man. Even the great prophet Nephi expressed sorrows similar to our own: "O wretched man that I am! Yea, *my heart sorroweth because of my flesh*; my soul grieveth because of mine iniquities. And when I desire to rejoice, *my heart groaneth because of my sins*" (2 Ne. 4:17, 19).

We're grateful that Nephi went on to explain how he overcame his negative feelings and got past the self-condemnation to allow his heart to rejoice and be glad for the redemptive powers of the gospel. "O then, if I have seen so great things," he said, "if the Lord in his condescension unto the children of men hath visited men in so much mercy, *why should my heart weep and my soul linger in the valley of sorrow*, and my flesh waste away, and my strength slacken, because of mine afflictions [for] I know in whom I have trusted" (2 Nephi 4:26, 19). This wonderful testimony teaches us that a cheerful heart is the result of two things: a mental *choice* of attitude and the correct *focus* of attention, for the principle is always the same: *What holds our attention holds us—both our thoughts and emotions.*

Of course our hearts become discouraged and we lose hope and resolve when we focus on outward circumstances that are shouting doom and gloom, instead of trusting in God to bring good out of even the darkest, most overwhelming opposition. "Hope deferred maketh the heart sick" (Prov. 13:12). When the Lord instructs us, "*Let* your hearts rejoice" (2 Ne. 9:52; see also D&C 100:12), he is revealing one of the keys to emotional control: granting ourselves *permission* to feel okay in spite of unwanted circumstances. He might have even said, "*Make* your hearts rejoice," because sometimes it requires not only our permission, but also a mental choice requiring effort and discipline to rout negative emotions and feel the joy we all should have in the knowledge of who we are as God's children and the destiny available to us through atonement and the plan of salvation. At these times, when joyful emotions are beyond our mortal reach, we can turn to the Lord and ask for help in letting our hearts rejoice, because he has the power to give us those feelings in our hearts. "And has he suffered that ye have begged in vain? Nay; he has poured out his Spirit upon you, and *has caused* that your hearts should be filled with joy" (Mosiah 4:20).

GOD KNOWS OUR HEARTS

"I know thy heart" (D&C 112:11).

Doctors can tell a lot about the condition of a person's *physical* health by taking a pulse or temperature reading. When the scriptures state "by their desires and their works you shall know them" (D&C 18:38), they're telling us that the values, desires, emotions, and passions we hold within our hearts are very much like a pulse or a temperature reading, indicating the condition of our inner *spiritual* health.

Just as a doctor uses a stethoscope to examine and judge the *physical* condition of our hearts, the Lord uses his divine discernment to search and ponder the *spiritual* condition of our hearts: "the Lord *looketh* on the heart" (1 Sam. 16:7), "the Lord *pondereth* the hearts" (Prov. 21:2), and "the Lord *searcheth* all hearts" (1 Chr. 28:9; see also

Jer. 17:10). This continual examination enables God to know the secrets of the heart, the thoughts and intents, the motives, dreams, desires, and all we treasure therein for the Lord "is a *discerner* of the thoughts and intents of the heart" (Heb. 4:12; D&C 33:1). Thus, always and ever, no matter how we might try to hide the truth from him, "God knoweth your hearts" (Luke 16:15).

Remembering that God monitors the emotional content and purity of our hearts should help us rise to a higher level of conduct. When people forget that God is constantly searching the content of their hearts, it's easy for them to rationalize the indulgence of their wicked desires as their deeds sink to a lower level, for *"they consider not in their hearts* that I remember all their wickedness"* (Hosea 7:2). But nothing in our hearts can remain hidden from God. "Shall not God search this out? *for he knoweth the secrets of the heart"* (Ps. 44:21) and "there is none to escape and there is no . . . heart that shall not be penetrated" (D&C 1:2).

If the realization that God knows the secrets of our hearts isn't enough motivation to repent and purify them, an additional motivation comes from realizing that the time will come when every unrepented secret will be known by our friends and families, for the time of judgment is coming in which he "will make manifest the counsels of the hearts" (1 Cor. 4:5) "that the thoughts of many hearts may be revealed" (Luke 2:35).

Not only does the Lord *examine* the condition of our hearts, but, like a coach who tests and tries the athlete's physical abilities, the Lord *tests and tries* the spiritual ability of our hearts, "for the righteous God *trieth* the hearts and reins" (Ps. 7:9; see also 1 Chr. 29:17; Prov. 17:3; 1 Thess. 2:4). And then, like a sports coach, the Lord will challenge us with loving tests and tutoring challenges because he loves us too much to simply stand idly by and watch us succumb to the lusts and passions of our flesh without reaching out to rescue us from ourselves.

We're familiar with the idea of this earth life as a proving ground during which, the Lord said, "we will prove them herewith, to see if they will do all things whatsoever the Lord their God shall command them" (Abraham 3:25). This probation concept was focused specifically on our hearts in the words of Moses, in which he reminded the Israelites how "God led thee these forty years in the wilderness, to humble thee, and to prove thee, *to know what was in thine heart,* whether thou wouldst keep his commandments, or not" (Deut. 8:2; see also 2 Chr. 32:31). "For the Lord your God *proveth you*, to know whether ye love the Lord your God with all your heart and with all your soul" (Deut. 13:3).

If a coach discovers a weakness in an athlete's performance, he immediately requires a program of corrective exercises. If a medical doctor discovers a heart problem during a physical exam, he urges immediate treatments to remedy the situation before it becomes life-threatening. When the Lord finds things in our hearts that are inappropriate, he, too, initiates correcting opportunities. Based on the values he finds treasured in our hearts, the Lord may say, "Thine heart is now right before me at this time" (D&C 39:8), or "there have been some few things in thine heart and with thee which I, the Lord, was not well pleased" (D&C 112:2). But a doctor or a coach can only *tell* us that we should exercise, stop smoking, or stay away from fattening foods. The Lord can *cause* the exact tests and tutoring circumstances that we need to come into our lives to challenge our growth so that our hearts learn to surrender the previously harmful desires and value the higher way. "I, the Lord, will contend with Zion, and plead with her strong ones, and chasten her until she overcomes and is clean before me" (D&C 90:36). "My son, despise not thou the chastening of the Lord, nor faint when thou art rebuked of him: For whom the Lord loveth he chasteneth, and scourgeth every son whom he receiveth" (Heb. 12:5-6). "No chastening for the present seemeth to be joyous, but grievous: nevertheless afterward it yieldeth the peaceable fruit of righteousness unto them which are exercised thereby" (Heb. 12:11).

GOD DWELLING IN OUR HEARTS

"The Lord hath said he dwelleth not in unholy temples, but *in the hearts of the righteous doth he dwell*" (Alma 34:36).

One important reason for accepting the Lord's corrective "spiritual exercises" and purifying our minds, hearts, and bodies as temples is to qualify ourselves for the companionship of the Holy Ghost and even the Savior himself. It's important to understand that when God finds a way past our barriers so that he can share his love with us, it isn't in the intellectual realm of the *mind* alone that he tries to catch our notice. He wants to touch us in the *heart,* the emotional center of our spiritual anatomy, where his presence will have the greatest effect upon our priorities, devotion, and actions. "Yea, it is the love of God, *which sheddeth itself abroad in the hearts* of the children of men" (1 Ne. 11:22; see also 4 Ne. 1:15).

When we receive comfort, or answers to prayers, these blessings are *recognized* mentally in the mind, but *felt* and *experienced* in the heart. "Thus, I will tell you in your mind and in your heart, by the Holy Ghost, which shall come upon you and *which shall dwell in your heart*" (D&C 8:2). "I now send upon you another Comforter," the Savior said, "that it may *abide in your hearts*" (D&C 88:3). "And behold, the Holy Spirit of God did come down from heaven, *and did enter into their hearts,* and they were filled as if with fire" (Hel. 5:45).

When God talks about "dwelling" in our hearts, he speaks of spiritual influence and companionship, not physical habitation, for "the idea that the Father and the Son dwell in a man's heart [physically] is an old sectarian notion, and is false" (D&C 130:3). When Christ came to earth the first time, there was no room for him in the inn. The question you and I face as we reach for his image and fellowship in our lives today is: Do we make room for him in our hearts? Paul expressed a desire for the disciples to experience such fellowship when he proclaimed the hope that Heavenly Father "would grant you, according to the riches of his glory, to be strengthened with might by his Spirit in the inner man; *That Christ may dwell in your hearts* by faith" (Eph. 3:16-17). "I will be on your right hand and on your left, and my *Spirit* shall be in your hearts, and mine angels round about you, to bear you up" (D&C 84:88).

WITH ALL THE HEART

The physical heart is divided into two sides with separate purposes and functions. Each side is divided into an upper chamber (the atrium) and a lower chamber (the ventricle). Four valves make sure that our blood can only travel in one direction through these chambers. (The sound we call the "heartbeat" is actually the sound of these valves slamming shut after each contraction and pulse of blood flow.)

The duty of the *left* side of the heart is to pump the blood outward to every cell in the body, providing oxygen and nourishment. The duty of the *right* side of the heart is to receive the blood back after the oxygen and nutrients have been delivered to the cells and then pump this depleted blood through the almost 500 square feet of lung surface, where carbon dioxide and other waste products are exchanged for new oxygen. It does this about twice a minute, every hour of every day. After the lungs have replenished the blood, it flows back to the left side of the heart to be sent on another journey around the body. This life-sustaining process is repeated more than a thousand times a day, with no conscious thought or effort on your part.

People can live fairly normal, healthy lives with only one lung or kidney, with only part of their stomach or intestines. But when it comes to the heart, it takes *all* of it working together to sustain life. Nothing less will do physically—or spiritually. Physically, *all* of the heart must do its part or the function will fail, our health will deteriorate, and physical death will result.

It's the same with our spiritual health. If our hearts fail to function with *full* commitment and integrity, our spiritual health deteriorates and results in spiritual death if the deficiency isn't corrected in time. We can't develop the spirituality that will prepare us for life in the celestial kingdom with a halfhearted effort any more than we could sustain physical life with a heart that only beat part-time instead of 24 hours a day. Trying to love and serve the Lord or trying to repent and obey with only *partial* yielding and commitment of heart is like expecting our physical heart to sustain life with only part of it being functional.

One of the most common themes in scripture is the need to love and serve the Lord with ALL the heart; with focused *singleness* of heart, real *intent* and *full purpose* of heart. This is our goal, but not where the natural man begins. As with all spiritual development, our heart's progress toward full and total commitment is accomplished step-by-step and line-upon-line. There is a specific process by which we grow from hearts that are indifferent or wavering, to hearts that have the strength and passionate dedication to the Lord's work that will take us to celestial rewards. Let's consider six steps in the process of growing into an "all-the-heart" level of discipleship: searching, accepting, turning, obeying, serving, and loving with all our hearts.

Step One: *Searching* **for Him "With All the Heart"**
We find and come to know the Lord by *searching* for him with all the sincerity and devotion of our hearts. "Come unto the Lord with all your heart," (Mormon 9:27) is the *invitation,* "And ye shall . . . find me, when ye shall search for me with all your heart" (Jer. 29:13) is the *promise.*

> But if from thence thou shalt seek the LORD thy God, thou shalt find him, *if thou seek him with all thy heart and with all thy soul* (Deut. 4:29).

Step Two: *Accepting* **What We Find "With All the Heart"**
The Lord said, "If men come unto me I will show unto them their weaknesses" (Ether 12:27). It's like helping us to see our true selves in a spiritual mirror. This discovery and self-awareness can be depressing, even overwhelming, unless we realize that discovering ourselves in his reflection is part of the divine process of opening the door to change from what we are now to becoming what he has in mind for us. "Only in Jesus Christ can any man learn the truth of what he is and how he can be changed from what he is to do the good for which he hopes" (*In His Footsteps Today* [Salt Lake City: Deseret Sunday School Union, 1969], p. 4). Satan taunts us with our weaknesses to discourage us, but Christ's love reveals them as a welcoming invitation. Finding ourselves by recognizing our personal deficiencies (the places we need to grow) is an important part of finding the Lord and learning his divine attributes, which we need to emulate in order to grow into his image.

"Don't blame yourself for being mortal and having imperfections," is the message, for "I *give* unto men weakness that they may be humble" (Ether 12:27). That may not seem like much of a welcoming response to our sincere effort to find and know him better, but discovering our weaknesses truly is a wonderful gift because the more we learn about God, the more we learn about ourselves and our need to grow more into his image. And the rest of this verse promises all who respond to this revelation of growth opportunities with humility and faith in Christ that he will add his divine power to theirs: "For if they humble themselves before me, and have faith in me, then will I make weak things become strong unto them" (Ether 12:27). It's through the process of first *recognizing* and then *exchanging* our deficiencies for his attributes that we can know him better because we've become more like him.

Step Three: *Turning* to Him "With All the Heart"

We face a decision as we look into the spiritual mirror Christ presented as we sought him with all our hearts. We face a turning point. We can either turn away from the newly revealed weaknesses he shows us as we retreat to our comfort zone, or we can "*turn* unto the Lord thy God *with all [our] heart*, and with all [our] soul" (Deut. 30:10). Halfhearted attempts to repent will never overcome the strength of bad habits. To reach fully for his image in our lives, we must "repent . . . and turn to the Lord [our] God with all [our] hearts and with all [our] might, mind, and strength," or we'll inevitably fall back into our old ways. (See D&C 98:47.) Let me share two examples.

I once taught the gospel to a friend who loved to smoke pipes. There came a time during the discussions when he had to choose between his physical pleasure and the Lord. During one of the lessons, he invited me to step outside with him. He took with him all his pipes and a brand-new canister of expensive tobacco. All these he placed on the sidewalk and proceeded to jump up and down on them until they were crushed and destroyed. This total commitment (with "all his heart," if you will) was a turning point that allowed the Holy Spirit to come into his life even more strongly than before. He became a devoted member of the Church. What would have been his experience if he had tried to just taper off something he loved so much instead of committing himself willingly, completely, with nothing held back?

In contrast, I have a dear sister who made all her own school clothes. She was really into the miniskirt fashions. One day when she was folding her clothes, I noticed her skirts had deep hems—six to eight inches. I asked her why. Grinning, she said, "Oh, that's so that someday, when I decide to repent, I won't have to make all new clothes." That kind of attitude is not what the Lord means by turning to him with all the heart.

As I watched this sister pass through the coming years filled with pain, tragedy, and heartache, outside of the Church more than in, I often wondered at the contrast between her heart and the heart of my pipe-smoking friend. Eventually she found the Lord and made her commitments to him, but how different a person's life is when he or she gives the heart to the Lord fully, holding nothing back, rather than clinging to it selfishly and protectively. "If I regard [value, prefer, and indulge] iniquity in my heart, the Lord will not hear me" (Ps. 66:18).

Step Four: *Obeying* Him "With All the Heart"

Having committed ourselves to turn wholeheartedly to the Lord, the next step is learn to keep his commandments with all our hearts. As we learn more about the Savior and ourselves, as we turn to him in repentance, the next requirement is to "keep and do" the commandments "with all thine heart, and with all thy soul" (Deut. 26:16), to "obey his voice . . . with all thine heart, and with all thy soul" (Deut. 30:2), and to improve the daily paths of our feet so that we "walk uprightly before [him] with all [our] hearts" (D&C 109:1).

Step Five: *Serving* "With All Our Heart"

Inward repentance to change *ourselves* is never enough in the Lord's kingdom, because genuine repentance also gives birth to the Christlike desire to help *others* change and find joy in the Lord as well. For example, the moment Enos gained realization of the miracle Christ had performed by removing the guilt from his heart, he wanted to share this good news with others. He said, "I began to feel a desire for the welfare of my brethren, the Nephites; wherefore I did pour out my whole soul unto God for them" (Enos 1:8).

Reaching out to bring others to Christ is a sure sign of a divinely changed heart. As Elder Merrill J. Bateman explained, "The person who is unenlightened *focuses inward* and is concerned with self, but once a person receives a spiritual witness and tastes the Savior's love, he or she is freed from constant introspection and concerns about self. By yielding to the Holy Spirit, such people may put off the natural man, turn their attention *outward*, and serve others" (*Ensign*, Jan. 1999, p. 10). "Therefore, O ye that embark in the service of God, see that ye *serve him with all your heart*, might, mind and strength, that ye may stand blameless before God at the last day" (D&C 4:2). Lukewarm and slothful service (without the dedication and devotion of *all* the heart) is unacceptable to the Lord. (See Rev. 3:16; D&C 58:29; 107:100.) "But take diligent heed to do the commandment and the law . . . to love the Lord your God, and to walk in all his ways, and to keep his commandments, and to cleave unto him, *and to serve him with all your heart* and with all your soul" (Josh. 22:5).

Step Six: *Loving* the Lord "With All Our Heart"

The Lord's intention in requiring "all of our hearts" in searching for him, repenting, obeying, and serving him is to produce not robots, but partners. Thus the goal of these activities is not like the bossy servitude between a soldier and his sergeant, or between an employee and his employer, but one of willing love and devotion that results in total commitment to Christ and his cause. "And thou shalt love the Lord thy God with all thy heart, and with all thy soul, and with all thy mind, and with all thy strength: this is the first commandment" (Mark 12:30). The Lord's goal is not to compel or manipulate us into obedience, but to win our support, obedience, and cooperation by convincing us that his way is best. He wants to teach us enough that we *agree* with him and want to do his will because we know it's the best choice we could make and because we love him so much it would hurt us not to please him.

> Wherefore, I give unto them a commandment, saying thus: Thou shalt love the Lord thy God with all thy heart, with all thy might, mind and strength; and in the name of Jesus Christ thou shalt serve him (D&C 59:5).

As we come to know, obey, and love the Lord with all our hearts, our devotion will grow to a compelling passion to *"cleave* unto [him] with all our heart" (D&C 11:19), to *"trust* in [him] with all [our] heart" (Prov. 3:5), and to *"praise* [him] with [our] whole heart" (Ps. 138:1). What these "all-the-heart" scriptures in this section are conveying is our need to devote ourselves to the Lord with "with all the energy of heart, that [we] may be filled with his love, which he hat bestowed upon all who are true followers of his son, Jesus Christ" (Moro. 7:48).

"Come unto me *with full purpose of heart,* and I will receive you" (3 Ne. 12:24).

Chapter Eleven

IN HIS IMAGE

"Therefore I would that ye should be perfect even as I, or your father who is in heaven is perfect" (3 Ne. 12:48).

We go to college to "become" a lawyer, a doctor, an engineer, a nurse, or some other type of professional. We come to this earth-school to become like Christ and Heavenly Father. The Savior asked, "Therefore, what manner of men ought ye to be?" (3 Ne. 27:27). His answer has compelling implications for the ways we must use our bodies. "Verily I say unto you, *even as I am*" (3 Ne. 27:27).

Because of our struggles against the fallen, carnal nature of our physical bodies, the only way we can look forward to attaining the perfection of his image is through the power of Jesus Christ, "who shall change our vile body, that it may be fashioned like unto his glorious body, according to the working whereby he is able even to subdue all things unto himself" (Philip. 3:21). This wonderful promise of transformation refers both to the future resurrection as well as the spiritual change of heart and nature we can each enjoy *right now*, in this life, as we are "sanctified by the Spirit unto the renewing of [our] bodies" (D&C 84:33). Through this divine intervention (or rescue) we can truly be "changed into the same [his] image from glory to glory, even as by the Spirit of the Lord" (2 Cor. 3:18). "The greatest miracle of the Atonement is the power Jesus Christ has to change our character if we come to Him with a broken heart and a contrite spirit" (Merrill J. Bateman, *Ensign*, Jan. 1999, pp. 12-13).

Through the gospel, the atonement of Christ, and the promises of spiritual rebirth "are given unto us *exceeding great and precious promises*" (2 Pet. 1:4) so that we can not only overcome our fallen, carnal nature and become new, worthy and acceptable in his sight, but also "that we may be *purified* even as he is pure" (Moro. 7:48), becoming "*pure and spotless* before God" (Alma 13:12), even "*partakers of his holiness*" (Heb. 12:10) and "*partakers of the divine nature*, having escaped the corruption that is in the world through lust" (2 Pet. 1:4). If we accept the spiritual rebirth and transformation into his image that he's striving to give us, then "we know that, when he shall appear, we shall be like him" (1 John 3:2; see also Moro. 7:48).

IS THIS REALLY POSSIBLE FOR ME?

Jesus tried to open our minds to the possibility of attaining his image, even while we're here in mortality, when he said, "I would that ye should be perfect even as I, or your Father who is in heaven is perfect" (3 Ne. 12:48). Because that's such a limitless and infinite target, we'll never reach it if we put mortal limitations of expectation on our ability to change. As we struggle against our fallen natures, we often doubt our ability to reach such lofty goals. However, we must nourish our faith and expectations with the assurance that "the Lord giveth no commandments unto the children of men, save he shall *prepare a way* for them that they may accomplish the thing which he hath commanded them" (1 Ne. 3:7).

When a painter paints a picture, it's confined within the size of the framework chosen to border the painting. All our lives we've heard slogans like, "If you think you can or you think you can't, you're right," or "Your attitude is more important than your aptitude." Like it or not, our attitudes and beliefs—in other words, the things we hold in the focus of our expectations—are interpreted by our minds as the *instructions* of what we want them to accomplish (or *not* accomplish). They form a mind-set (like the four-minute mile) that forms the boundaries of our possibilities and the limits of the things we are giving our brains permission for us to experience or achieve. The reason motivators say things like "Lengthen your stride" or "Reach for the stars" is to open our minds to possibilities that have no unnecessary borders or limitations.

One of Satan's most frequent lies is that we're stuck in our faults and we can't change. The issue is simple: God says that with his help we *can* change every imperfection, no matter how long they've held us in captivity, and Satan says *we can't*. The gospel message is that because of the Lord's power (and willingness) to change us, there is no desire, no habit, no addiction, no part of our fallen nature so deeply ingrained or permanent that it's inescapable, because his grace is always greater than our need. As President Benson testified, "There is no human problem beyond His capacity to solve. Because He descended below all things (see D&C 122:8), He knows how to help us rise above our daily difficulties . . . there is no evil which He cannot arrest" (*Ensign*, Nov. 1983, p. 8).

Because of our fallen nature, it's very easy to tell ourselves, "I can't help it; it's just the way I am." This rationalization may be soothing to a mistreated conscience, but it's not conducive to receiving "the mighty change of heart." Our favorite weaknesses and sins often drag on and on because, in spite of our feelings of guilt, we're somehow more comfortable with the pain they cause than we are with the idea of changing and growing. Satan has ridiculed the Savior and tried to persuade mankind that "the doctrine of Christ [is] a foolish and a vain thing" (3 Ne. 2:2). When we doubt the Savior's power to change us we probably don't use those same words, but don't we mean the same thing when we conclude that our problems are just too hard for the Lord?

When we doubt ourselves, we're really doubting the Savior, who promised, "My grace is sufficient" (Ether 12:27). To believe that we can't change into his image and escape our inappropriate behavior is to deny everything God has promised—and that he has already done in the lives of countless thousands who have changed carnal thoughts and behaviors to spiritual ones.

We believe that Jesus changed the water into wine, so why are we so hesitant to believe in his power to change us? He magnified a few loaves of bread and fishes to meet the needs of thousands. Why, then, is it so hard to believe that he can magnify our efforts until we gain the victory over our fallen nature? (See John 2, Mark 6.)

Here your most crucial challenge, once recognizing the seriousness of your

mistakes, will be to believe that you can change, that there can be a different you. To disbelieve that, is clearly a Satanic device designed to discourage and defeat you. . . . Only he would say, "You can't change. You won't
change. It's too long and too hard to change. Give up. Give in. Don't repent.
You are just the way you are." That, my friends, is a lie born of desperation.
Don't fall for it (Jeffrey R. Holland, *The New Era,* Oct. 1980, pp. 11-12).

WE CAN'T—BUT HE CAN

A woman who was working her way out of difficult circumstances
once said to me, "I felt so alone and helpless. I thought that because I
had gotten myself into this situation, I didn't have the right to ask for
help to get out of it."

Haven't we all had that feeling? And that's exactly how Satan wants
us to feel so that we won't draw upon the Savior's power. She continued, "The thing that helped me the most is learning that I can't do it
alone. It's such a relief to know that I'm not expected to." The act of
will that turns us toward him in repentance entitles us to his help,
even *before* we have the strength to keep our covenants flawlessly. His
grace and redeeming power is a *partnership* in our struggles more than
a *reward* at the end of the struggle, for they will never end if we insist
on fighting the battles against our fallen nature alone.

Trying to change ourselves by will power and self-control alone is
about as effective as shooting at a battleship with a BB gun. No matter how sincere we are, no matter how earnestly we try, *we cannot*
change the carnal nature of our fallen, natural-man hearts and dispositions by our own efforts. We may do much good in *controlling* our
behavior, but until we allow Jesus Christ to alter our hearts and
desires, we'll continue to suffer the constant struggle between the
desires of the fallen flesh and the will of the spirit.

This may be one of the most difficult and bitter lessons we mortals
must learn. It would be ludicrous to imagine a heart patient daring to cut
his own chest open and attempt to perform his own heart transplant. Yet,
in the arrogance of our mortal pride, this is exactly what many of us try
to do spiritually. "Who can say, I have made my heart clean, I am pure
from my sin?" (Prov. 20:9). Only Christ can do that, for spiritual death
can never bring itself back to life. "If we rely on *our* intellect and reason
[alone] we can only go as far as the rest of mankind" (*Ensign,* June 1988,

p. 19). "*Only in him* can any man find the strength, the power and the ability to live a godly life. Only in Christ is there power to transform the human mind and the human heart" (*In His Footsteps Today* [Salt Lake City: Deseret Sunday School Union, 1969], p. 4).

Our covenants and commitments, our goals and resolves, our will power and self-discipline, our most valiant efforts to restrain our evil habits and desires are all necessary and appreciated, but are insufficient to save us without his transforming power. Going to the celestial kingdom isn't based on controlling our evil desires with superhuman restraint and will power. The evil is only caged and locked inside us like a ticking time bomb, just waiting for the right temptation to light the fuse. What's required is a *complete transformation* of the heart, desires, and fallen human nature. It's the difference between our merely *controlling* our bad habits and allowing Christ to change our hearts and natures so that we no longer *want* the sins. And nothing but the blood, atonement, and grace of Jesus Christ can do that. As President Ezra Taft Benson emphasized, "Only Jesus Christ is uniquely qualified to provide that hope, that confidence, and that strength to overcome the world and rise above our human failings" (*Ensign*, Nov. 1983, p. 6).

Christ came to rescue us from our sins, not to sit idly by, waiting until we have somehow cleansed and delivered ourselves. He is the Savior; we are not. He is the Redeemer; we are not. He is the only way back to the Father. He said, "I will fight your battles" (D&C 105:14). "I will also save you from all your uncleanness" (Ezek. 36:29) and "I am able to make you holy" (D&C 60:7).

But just as changing or lifting our own fallen nature to the spiritual level God wants to give us isn't something we can do for ourselves, neither is it something God can do for us without our permission and cooperation. But when the two of us work together as partners, it's possible. As the Savior said, "As the branch cannot bear fruit of itself, except it abide in the vine; no more can ye, except ye abide in me. I am the vine, ye are the branches: He that abideth in me, and I in him, the same bringeth forth much fruit: for without me ye can do nothing" (John 15:4–5). "Yea, I know that I am nothing; as to my strength I am weak . . . but . . . in his strength I can do all things" (Alma 26:12; see also Philip. 4:13).

The Savior taught that, at the day of judgment, each of our lives will be reviewed to see what we have done for him, in response to what he has done for us. (See Matt. 25:31-46 and Rev. 20:12.) But I think we should also expect to be judged by what we have allowed Jesus Christ to do for us. The Savior spoke often of the importance of a broken heart and a contrite spirit, and of the need to rely upon him and his power. I expect our lives will be examined to see if we were willing to admit our need for him and throw ourselves upon his grace and mercy, or if we allowed our pride to keep us struggling throughout life with our self-sufficiency and will power getting in the way of his power to make us new.

He has prepared to transform us into the holiness and perfection of his image by helping us to be spiritually reborn—to receive a new heart, a new mind, indeed, an entirely new nature and disposition.

BORN AGAIN

"And now behold, I ask of you, my brethren of the church, *have ye spiritually been born of God?* Have ye received his image in your countenances? Have ye experienced this mighty change in your hearts?" (Alma 5:14).

Scriptures that discuss the concept of being born again always present it as a spiritual *necessity*, not as a suggestion. For example, Jesus said, "Except a man be born again, he cannot see the kingdom of God," and then counseled his confused audience to "Marvel not that I said unto thee, Ye *must* be born again" (John 3:3, 7). Alma emphasized the same theme: "I say unto you the aged, and also the middle aged, and the rising generation . . . that they *must* repent and be born again" (Alma 5:49). Again: "Now I say unto you that ye must repent, and be born again; for the Spirit saith *if ye are not born again* ye cannot inherit the kingdom of heaven" (Alma 7:14).

What does it mean to be "born again"? Dictionaries define being "born" as being brought into life or existence. This isn't entirely accurate. It's true that when a baby is born it has just been brought into a new existence in mortality. It's new in *this* sphere of existence and experience. But we know that we, as spirit identities, existed as individual entities in the premortal world of spirits. Our mortal birth was merely a *transfer* of our spirit-selves from that spirit world in heaven into this temporary world to experience a new physical life or existence. (In this sense even death could be termed a birth, for it's nothing more than the same spirit identity leaving this physical world and entering into a new life or existence in the spirit world to await the resurrection.)

Paul emphasized the born-again experience when he stated that "if any man be in Christ, he [becomes] a *new creature*: old things are passed away; behold, all things are become new" (2 Cor. 5:17). He emphasized our need to be changed by the Spirit when he stressed our need to "*put off the old man* with his deeds," meaning setting aside the natural man ways of the flesh and to "*put on the new man*, which after God is created in righteousness and true holiness" (Eph. 4:24). The Book of Mormon clarifies such statements to mean being "*changed* from their carnal and fallen state, to a state of righteousness, being *redeemed* of God, becoming his sons and daughters; And *thus they become new creatures*" (Mosiah 27:25). We can see, then, that being "born again" spiritually means being elevated from a merely physical, or natural-man and fallen, carnal sphere of existence into a new life and existence where we become aware of and then experience a higher level of existence—a life of spirituality.

King Benjamin told his converts, whose hearts, nature, and actual dispositions had been changed by faith in Christ, that "ye shall be called the children of Christ, his sons, and his daughters; for behold, this day *he hath spiritually begotten you*; for ye say that your hearts are changed through faith on his name; therefore, ye are born of him and have become his sons and his daughters" (Mosiah 5:7). Prophets describe this gift as a "mighty change" because of the dramatic, freeing effect it has on one's struggle with the heart's natural carnal desires. Being born again and becoming a new creature in Christ is like having the natural, inborn compulsions to sin removed, so that instead of being constantly compelled by the flesh to do things we know are inappropriate, we're freer to choose the higher way. Of course, we always had the *agency* to make those higher choices, but not always the *ability*, because of the preoccupation the unchanged fallen man has with satisfying the flesh and choosing the paths of gratification.

Those who consider the phrase "born again" to be a Protestant term might be interested to know that this term occurs more frequently (and receives greater textual dissertation) in LDS scripture than it does in the Bible alone. For example, after Alma's "born again" experience, he testified, "I have repented of my sins, and have been redeemed of the Lord; behold *I am born of the Spirit*" (Mosiah 27:24). He then testified of our need to be born again as well: "And the Lord said unto me: Marvel not that *all mankind,* yea, men and women, all nations, kindreds, tongues and people, *must* be born again" (Mosiah 27:25). An essential part of this spiritual rebirth and renewal is the process of receiving a new heart and a new mind.

SPIRITUAL HEART TRANSPLANTS

"Thou shalt offer a sacrifice unto the Lord thy God in righteousness, even that of a broken heart and a contrite spirit" (D&C 59:8; see also Ps. 51:17; 3 Ne. 9:20).

Normally we feel bad when something is broken because it has lost its usefulness. It's a strange contrast that while the Lord asks us to do our best to fix all the rest of the imperfections in our lives, there is one part he actually wants *us* to break so that *he* can repair it. You must "come unto me," he said, "with *a broken heart* and a contrite spirit" (3 Ne. 12:19). Offering him the sacrifice of a heart that is broken with sorrow because of our sinfulness is so important that he has warned of serious consequences to those "whose hearts are not broken, whose spirits are not contrite" (D&C 56:17). For example, when we try to protect our hearts by refusing to acknowledge our imperfections instead of offering them to Christ to be changed, we deny ourselves the blessings of the Atonement and put ourselves outside its power, because "Jesus was crucified by sinful men for . . . the remission of sins unto the contrite heart" (D&C 21:9). "Behold, he offereth himself a sacrifice for sin, to answer the ends of the law, unto all those who have a broken and a contrite spirit; and unto none else can the ends of the law be answered" (2 Ne. 2:7).

When a person's physical heart can no longer function well enough to support life, the patient either dies or is rescued by a heart transplant. In major medical centers across America, thousands of heart patients wait for some stranger to die and give them a second chance at life. They're waiting anxiously for a heart that matches their physical size, blood, and tissue type that will be suitable for transplant.

Every one of these patients is on a countdown toward death. If an individual's heart condition is severe enough to warrant a transplant, it's only a matter of time before the heart grows so weak that it can't pump enough blood to keep the person alive. The patient knows he or she will either leave the hospital with a new heart or die waiting. This presents heart doctors with a terrible dilemma. They don't want to put people on the transplant list until they're so sick that nothing else will keep them alive, but if they wait too long, the patients will be too sick and weak to survive the traumatic surgery.

And there are never enough hearts. Each year about 15,000 Americans are potential candidates for heart transplants, but only about 2,000 hearts are donated. For every patient who receives a heart transplant, at least two more join the wait.

In Chapter Ten we learned that every mortal descendent of Adam has a defective, fallen heart that is "deceitful above all things, and desperately wicked" (Jer. 17:9), and that leads mankind's passions more naturally toward selfishness and carnality than toward selflessness and spirituality. As the Master Physician, Christ proclaimed that one of the major purposes of his ministry is "*to heal the brokenhearted*" (Luke 4:18). He certainly has the ability to "*revive* the heart of the contrite ones" (Isa. 57:15), to "*strengthen* [our] heart" (Ps. 27:14), and to "*enlarge*" our hearts (Ps. 119:32). But his real goal is not to patch us up, but to give a *new heart* to every person who will accept it. While there are never enough physical hearts donated to supply the need for transplants, there are no such limitations on the Lord's ability to change or replace hearts spiritually. "I will *give them* an heart to know me," he said, so "they shall be my people, and I will be their God" and so that "they shall return unto me with their whole heart" (Jer. 24:7). "*A new heart also will I give you*, and a new spirit will I put within you: and I will take away the stony heart out of your flesh, and I will give you an heart of flesh" (Ezek. 36:26; see also 11:19).

Jesus Christ is the greatest physician who ever lived and healed upon this planet. Demonic possession, twisted and crippled limbs, blinded eyes, deaf ears, palsy, leprosy—it mattered not, for there was no problem too great for the power of his tender touch of healing love. As great as those physical healings were, however, none are so awesome and precious as his tender and healing touch upon the heart of a sin-sick soul. Somehow the irresistible power of his enduring love can reach into the heart and soul of a lost and despairing sinner and heal every spiritual wound.

In some divine way we don't fully understand, he still works that miracle today, transforming the wretched, fallen heart so that each of us can become a new creature, a beautiful child of God, born anew to the image of our Lord and Savior.

Receiving a new heart spiritually removes the power of the carnal flesh—the addictions and natural-man traits that previously held us captive to improper desires—and restores the spiritual ability to use our agency to follow the Lord. The Lord's purpose in giving us a new heart is "that he may *incline our hearts* unto him, to walk in all his ways, and to keep his commandments, and his statutes, and his judgments, which he commanded our fathers" (1 Kgs. 8:58). The blessings we receive are just not the same when they come from torturous, iron-jawed willpower and force-of-mind compliance as when our obedience comes from the willing and true desires of our heart. "I, the Lord, *require* the hearts of the children of men" (D&C 64:22) means that he wants us to yield our hearts to him to be changed, or even replaced, so that we can cooperate with him with *"full purpose of heart,* acting no hypocrisy and no deception before God, but with real intent," so that our obedience is eager and willing (2 Ne. 31:13).

Even if we could obey all the commandments by willpower alone, it wouldn't be real or sufficient. Unless our hearts are in it, it couldn't change or sanctify us. For example, Israel's king, Amaziah, was a man who "did that which was right in the sight of the Lord" *outwardly,* "but not with a perfect heart" (2 Chr. 25:2). This will never do, for what is required is a divine intervention to lift us above the desires and preferences of the fallen, natural man, the giving of new, higher, more spiritual desires, a new disposition or nature. People who have experienced a spiritual heart transplant have testified that "the Spirit of the Lord Omnipotent . . . has wrought a mighty change in us, or in our hearts, that *we have no more disposition to do evil,* but to do good continually" (Mosiah 5:2). Ammon's converts declared the same encouraging testimony "that their hearts had been changed; that *they had no more desire to do evil*" (Alma 19:33). It's like waking up from a trance where your mind and heart had been continually occupied with carnal things of the flesh and discovering a whole new world of spiritual joy and freedom you never even knew existed. "Behold, he changed their hearts; yea, he awakened them out of a deep sleep, and they awoke unto God" (Alma 5:7). "And behold, he preached the word unto your fathers, and *a mighty change was also wrought in their hearts,* and they humbled themselves and put their trust in the true and living God. And behold, they were faithful unto the end; therefore they were saved" (Alma 5:13).

With today's high-tech medicine, the success rate for heart transplants is about 95%! Yet, within five years, 40% of these people will die anyway, and within six to eight years, almost every one of them will be dead. So what is the Lord's rate of success when he changes a heart? It depends on the person's agency. Alma, whose life was rescued by such a mighty change of heart, frequently challenged the members of the church not to mistake *membership* for spiritual *transformation.* "And now behold, I ask of you, my brethren of the church, have ye spiritually been born of God? Have ye received his image in your countenances? Have ye experienced this mighty change in your hearts? And now behold, I say unto you, my brethren, if ye have experienced a change of heart, and if ye have felt to sing the song of redeeming love, I would ask, *can ye feel so now?*" (Alma 5:14, 26). "He healeth the broken in heart, and bindeth up their wounds" (Ps. 147:3).

RECEIVING THE MIND OF CHRIST

One of the interesting stories in the Apostle Paul's ministry occurred when he was shipwrecked on the island called Melita. Cold and wet from swimming to shore, Paul had helped the survivors gather wood for a fire when he was bitten by a deadly snake. "And when Paul had gathered a bundle of sticks, and laid them on the fire, there came a viper out of the heat, and fastened on his hand. And when the [superstitious] barbarians saw the venomous beast hang on his hand, they said among themselves, No doubt this man is a murderer, who, though he hath escaped the sea, yet vengeance suffereth not to live." But Paul was not concerned about the bite from the poisonous snake and merely "shook off the beast into the fire, and felt no harm." The natives knew the deadly snake well and expected his immediate death, so "they looked when he should have swollen, or fallen down dead suddenly: but after they had looked a great while, and saw no harm come to him, *they changed their minds,* and said that he was a god." (See Acts 28:1-6.) As we reach for the image of Christ in our lives, we all need to "change our minds" so that we learn to think more spiritually, more purely, more selflessly, more productively, and more like Christ.

The Lord has invited all of us to be changed into his image and return, with him, to live forever in the celestial kingdom with our heavenly parents. Yet he said that "there are many called, but few are chosen. And why are they not chosen? Because their hearts are set so much upon the things of this world" (D&C 121:34-35). "Set your affection on things above, not on things on the earth" (Col. 3:2).

Mortals can have many mind-sets or "frames of mind" besides one that is conducive to the Lord's spirit and to mental renewal or transformation. For example, people can be career-minded, music-minded, family-minded, or church-minded. Each of these states of mind can be beneficial or harmful, depending on how much of our time they occupy and how we apply them. For example, while sports are not evil in and of themselves, we know people who are so sports-minded that it preoccupies almost every waking thought and emotion. The scriptures indicate that our main goal in developing our minds is to "have the mind of Christ" (1 Cor. 2:16). Because this world is filled with so many things to occupy our minds, attention, and emotions, Christ has asked us to avoid worldly distractions and to "look unto me in every thought" (D&C 6:36). Such a focus of attention doesn't come naturally to the fallen man.

In Chapter Nine we learned that one consequence of the fall of man (which we inherited from Adam and Eve) was the corruption of our minds, causing our thoughts and imaginations to tend toward evil instead of righteousness, and to become carnal and in a state of "enmity against God," all of which prevents us from abiding in his presence or enjoying the companionship of his Holy Spirit. (See 2 Tim. 3:8; Gen. 6:5; Ne. 8:7; and D&C 67:12-13.) But this consequence doesn't mean that we're stuck with fallen thought-patterns. The mind can be changed, it can be trained, it can be expanded and enlarged, and, through the power of Christ, it can be renewed and brought into full harmony with Christ and the Holy Spirit.

Regarding our quest for Christ's image, Paul said, "Let this mind be in you, which was also in Christ Jesus" (Philip. 2:5). Paul would never make such a challenge to our thought process lightly, for he was keenly aware of the natural and carnal mind tendencies we all face. But he gave the key to attaining the mind of Christ when he said that we all need to be "*renewed* in the spirit of your mind" (Eph. 4:23). He explained further: "And be not conformed to this world: but be ye *transformed* by the *renewing of your mind*, that ye may prove [live up to and fulfill] what is that good, and acceptable, and perfect, will of God" (Rom. 12:2). Paul also affirms that when we put Christ and his promises at the center of our attention, when we allow him to change our heart and nature, he can renew our minds in such a way that we can not only destroy the "strong holds" of evil habits and addictions, but even "cast down [evil] imaginations" and bring *every thought* into harmony with Christ! "For though we walk in the flesh, we do not war after the flesh: (For the weapons of our warfare are not carnal, but mighty through God to the *pulling down of strong holds*;) *Casting down imaginations*, and every high thing that exalteth itself against the knowledge of God, and bringing into captivity *every thought* to the obedience of Christ" (2 Cor. 10:3-5).

Another part of the process of spiritually renewing our minds is to have our mental capacities expanded so that we perceive and comprehend things that we previously couldn't understand, or perhaps didn't even notice. "The natural man receiveth not the things of the Spirit of God: for they are foolishness unto him: neither can he know them, because they are *spiritually discerned*" (1 Cor. 2:14). The thirty-second chapter of Alma presents a wonderful treatise on spiritual mind expansion and how to allow God's truths and Spirit to enter one's mind and heart so that "your understanding doth begin to be enlightened, and your mind doth begin to expand" (Alma 32:34). Joseph Smith described that mental expansion after he and Oliver Cowdery were baptized, and they "were filled with the Holy Ghost," with the result that "Our minds *being now enlightened*, we began to have the scriptures laid open to our understandings, and the true meaning and intention of their mysterious passages revealed unto us in a manner which we *never could attain to previously*, nor *ever before had thought of*" (JS-H 1:73-74).

PREREQUISITES

Receiving a spiritual heart transplant and a mind spiritually renewed requires a cooperation between God and man, for it is "faith and repentance [that] bringeth a change of heart unto them" (Helaman 15:7). "And *according to his faith* there was a mighty change wrought in his heart" (Alma 5:12; see also Mos. 5:7).

Just as a doctor asks his patient to do certain things to prepare his body for a heart transplant, there are some specific things the Lord has asked us to do to make his divine intervention possible in fulfilling the "exceeding great and precious promises" (2 Pet. 1:4) of a spiritual rebirth—a new heart, mind, and disposition. Four of these essential preparations are to commit ourselves to his standard of cleanliness, holiness, and purity; to commit ourselves to stand in holy places, separate from the world; to do all we can to rise above our fallen natures; and then to surrender the rest to him.

1. Mental and Emotional Purity
"Be ye holy; for I am holy" (1 Pet. 1:15).

In this mortal world, it's constantly necessary to clean things. We wash our dishes, our clothing, our bodies. We must also learn to clean our minds and spirits. Repentance is ever urged: "Wash you, make you clean; put away the evil of your doings from before mine eyes; cease to do evil" (Isa. 1:16). How foolish and unhealthy it would be, when we wash our dishes, if we only cleaned the *outside* of the cup, the glass or pan, and ignored the germs and dirtiness on the inside, where we place our food. "And the Lord said unto him, Now do ye Pharisees make clean the outside of the cup and the platter: but your inward part is full of ravening and wickedness" (Luke 11:39). The purpose of the gospel transformation is not to make us look good on the *outside*, but to give us genuine changes and healing on the *inside*.

I once worked in a business where the man in charge of hiring had an interesting approach to selecting new employees. No matter how nicely dressed the applicants were, before he would grant an interview he made an inspection of their cars. If an applicant's car was clean inside and out, he granted an interview, but if the car was dirty on the outside or cluttered inside, the person didn't even get an interview.

You may feel that practice was unfair or discriminatory, but it does illustrate that it's always easier to clean the outside of our bodies, to wear nice clothes, and *appear* presentable than it is to cleanse and keep ourselves pure and clean on the inside where the real action is and where eternal consequences are being determined. Jesus spoke often of this hypocrisy: "Woe unto you, scribes and pharisees, hypocrites! For ye are like unto whited sepulchers, which indeed appear beautiful outward, but are within full of dead men's bones, and of all uncleanness. Even so ye also outwardly appear righteous unto men, but within ye are full of hypocrisy and iniquity" (Matt. 23:27-28).

Our Heavenly Father dwells in an environment of perfect purity and holiness, "and no unclean thing can enter into his kingdom," for "no unclean thing can dwell there, or dwell in his presence" (3 Ne. 27:19; Moses 6:57). Thus Paul emphasized that "denying ungodliness and worldly lusts, we should live soberly, righteously, and godly in this present world" so that we become "holy and without blame before him in love" (Titus 2:12; Eph. 1:4). Moroni echoed the same standard when he challenged his readers "that ye become holy, without spot" (Moro. 10:33). The scriptures speak of our need to have a "pure conscience" and a "pure heart." (See 1 Tim. 3:9; 1:5 Ps. 24:4; 2 Ne. 25:16; D&C 41:11.) The question we must ask ourselves is this: If the Lord were reporting to Heavenly Father about our lives and the choices we're making for our bodies, could he use those two terms to describe the condition of our spiritual anatomy? Could he tell our Father that we've done our best to meet the requirements for receiving the mighty change? "Having therefore these promises, dearly beloved, let us cleanse ourselves from all filthiness of the flesh and spirit, perfecting holiness in the fear of God" (2 Cor. 7:1).

2. Be Ye Separate

"And now I say unto you, all you that are desirous to follow the voice of the good shepherd, come ye out from the wicked, and be ye separate, and touch not their unclean things" (Alma 5:57).

Satan has filled this world with clever imitations of everything that is righteous and holy: deceptive, alluring substitutes that mimic the clean and pure but drag us down to sin and defeat. One of mortal probation's challenges is to learn to recognize the "difference between holy and unholy, and between unclean and clean" (Lev. 10:10; see also Ezek. 44:23). Of course, merely *recognizing* the difference between Satan's substitutes and divine purity is not enough. We must also *choose* the spiritual path of cleanliness and purity if we are to receive the mighty change that makes us more like our Savior and Heavenly Father. Thus Paul counseled, "be not conformed to this world" (Rom. 12:2) and the Lord said, "I will raise up unto myself a pure people, that will serve me in righteousness" (D&C 100:16). James said that "pure religion and undefiled before God and the Father is this . . . to keep himself *unspotted from the world*" and that "whosoever therefore will be a friend of the world is the enemy of God" (James 1:27; 4:4). "Love not the world, neither the things that are in the world. If any man love the world, the love of the father is not in him" (1 John 2:15-17).

3. Subdue and Have Dominion

"We have a labor to perform whilst in this tabernacle of clay, that we may conquer the enemy of all righteousness, and rest our souls in the kingdom of God" (Moro. 9:6).

To prepare ourselves for the new birth Christ is seeking to give us, we must not only stand in holy places, separating ourselves from the world's values, but we must also commit ourselves to do whatever it takes, for as long as it takes, to rise above and conquer the weaknesses of our fallen nature. God did not place us in fallen bodies to become prisoners to their unworthy passions, but to learn to bring them into subjection to the will of the spirit. "And God blessed them, and God said unto them . . . replenish the earth and *subdue it: and have dominion*" (Gen. 1:28). To "subdue" and "have dominion" over our bodies means to control *them* rather than being controlled *by* them. "I beseech you therefore, brethren, by the mercies of God, that ye present your bodies a living sacrifice, holy, acceptable unto God, which is your reasonable service" (Rom. 12:1). David O. McKay said, "There can be no fullness of life where there is slavery; and the man who is subject to his appetite and passions is the most abject slave. He who can rule his passions is greater than a king." He then emphasized that a "guiding principle to a realization of a higher life is the power of self-denial and the resultant self-mastery" (*Pathways To Happiness* [Salt Lake City: Bookcraft, Inc., 1957], pp. 89-90).

4. Yielding to Him

". . . even to the purifying and the sanctification of their hearts, which sanctification cometh *because of their yielding* their hearts unto God" (Hel. 3:35).

Something interesting happened to the prophet Jeremiah that can help us understand the part "yielding" has to receiving a change of nature. He tells us that the word of the Lord came to him, saying,

"Arise, and go down to the potter's house, and there I will cause thee to hear my words" (Jer. 18:2). We may be sure that Jeremiah's curiosity was aroused by this strange instruction. Certainly, he was used to communing with the Lord on isolated mountain peaks or in the barren wilderness, but why in a cluttered, dusty potter's shop?

Jeremiah tells us that when he arrived, the potter was busy at his trade, "and, behold, he wrought [made] a work on the wheels" (Jer. 18:3). As Jeremiah watched him form the clay into a beautiful vessel on the flat spinning wheel, something quite natural happened, something which the Lord knew was going to happen. "And the vessel that he made of clay was marred in the hand of the potter: so he made it again another vessel, as seemed good to the potter to make it" (Jer. 18:4).

As Jeremiah watched the potter's careful work develop a flaw that made the vessel unacceptable and unusable, he must have sensed the potter's disappointment. After all that work, the vessel was ruined. But, to Jeremiah's surprise, the potter didn't discard the broken vessel for a new lump of clay. Patiently, the potter crushed the vessel back into a lump of clay. After carefully kneading it, he put the very same clay back onto the spinning wheel and began to refashion the clay that had been "marred" into the vessel he had in mind for it all along. There was no anger. The marred clay was not cast aside for a new lump. The potter simply accepted the flaw that appeared as a natural part of life and patiently began again to achieve his objective. As Jeremiah watched this happen, as he witnessed this wonderful demonstration of tolerance, patience, and even forgiveness, the word of the Lord came to him again, saying: "O house of Israel, cannot I do with you as this potter? Saith the Lord. Behold, as the clay is in the potter's hand, so are ye in mine hand, O house of Israel" (Jer. 18:6).

And then Jeremiah, who had suffered so much disappointment and frustration over Israel's stubborn refusal to submit their lives to the will of the great Jehovah, suddenly understood. Israel's apostasy and sin could not discourage the Lord! God's love for us is stronger than our sin! Nothing we do can cause Him to reject us. No matter how great our flaws, no matter how many times he has to forgive and start over, God is going to persist in his love and patience towards us until all who will respond to him are able to live the commandments and become the people, the very Zion, he is trying to establish.

Now Jeremiah began to comprehend the reality of God's amazing forgiveness and patience with a faltering Israel. He finally understood that God sees our flaws as temporary problems, as opportunities for us to grow and become better. Now Jeremiah understood the wonderful truth that God will always see past our sins and weaknesses to our divine potential. He knows that every one of us can achieve this potential if we only yield ourselves to his will, like clay in the potter's hand, and follow the commandments.

We may need to learn many lessons before we're prepared to receive a spiritual heart transplant. King Benjamin stressed that one of the prerequisites to becoming a new creature in Christ and passing from the status of being a natural man (and enemy to God) is the essential process of *yielding* "to the enticings of the Holy Spirit." This means yielding willingly and appreciatively to whatever is required, in the wisdom of the Lord, to teach and train and perfect us. Yielding to the Holy Spirit, said King Benjamin, means being "willing to submit to all things which the Lord seeth fit to inflict upon [us], even as a child doth submit to his father" (Mosiah 3:19) or as the soft clay yields to the wisdom and skill of the potter. For "the purifying and the sanctification of their hearts . . . cometh *because of their yielding* their hearts unto God" (Hel. 3:35).

But there's more to "yielding" than simply being submissive to the adversities and tutoring circumstances the Lord may use to prepare us for the mighty change. If we would be changed by Christ, we must not only yield to him our will in obedience and covenants, but also surrender to him our very sins, our faults, our bad habits, weaknesses, hurts, fears, doubts – everything rotten and unworthy of one who belongs to him. Only then can he remove all these spiritual infections and cleanse our body and spirit. How could he possibly make us new if we insist on bringing with us all the old failures?

Every year in America, approximately 30,000 people commit suicide. At least another 100,000 attempt to take their own lives. But millions of us are slowly committing spiritual suicide as we humiliate, punish, and generally diminish ourselves with negative and false beliefs about our fallen natures and past mistakes. Part of "yielding" to Christ means letting go of every grudge, every feeling of bitterness, and every bit of malice, whether toward ourselves or toward others. If we want the Lord to make us new, we must be willing to abandon not only the sins and weaknesses diminishing our spirituality, but also our inappropriate memories. "If thou return to the Almighty . . . thou shalt put away iniquity far from thy tabernacles" (Job 22:23). "*Cast away* from you all your transgressions, whereby ye have transgressed; and make you a new heart and a new spirit" (Ezek. 18:31). "Casting" our transgressions away from us means not only repentance and improved behavior, but also removal of those transgressions from our memories and habits of self-abasement.

HOW LONG WILL IT TAKE?

Paul desired "that he would grant you, according to the riches of his glory, to be strengthened with might by his Spirit *in the inner man*" (Eph. 3:16). Overcoming the desires of the natural man isn't easy, and in most cases it doesn't come quickly. Virtue and holiness do not come into our lives from one single choice. The battle between the righteous will of our spirit and the desires of our flesh is constant. To conquer the flesh requires a lifetime of persistent effort and determination. It often involves making mistakes, stumbling, and renewing our resolves and efforts. Thus Christ has said, "ye must *practice* virtue and holiness before me continually" (D&C 46:33; 38:24).

It takes between six and ten hours for a team of highly skilled surgeons to perform a heart transplant. How long does it take the Lord to change our hearts spiritually and give us a new nature and disposition? Depending on our cooperation, a spiritual heart transplant may occur almost in an instant, as with Paul or Alma, or take years to accomplish. For most of us, receiving the mighty change of a new heart and a renewed mind is not a miraculous *event* but a more normal *process* of gradual improvement and sanctification. It's a precious gift that we must then work to make permanent in our characters and personalities as Christ, who changed our spirit's desires and disposition, also works to change our nature and help us bring our bodies and physical actions into harmony with our spiritual state. As Elder Merrill J. Bateman said:

> Few mortals share with Alma the Younger or Paul the Apostle the dramatic experiences which resulted in their spiritual rebirths over short periods of time. In fact, I believe those experiences are recorded in the scrip-

tures not to define the time frame during which one may be reborn but to provide a vivid picture of what the *accumulated, subtle changes are that take place in a faithful person over a lifetime* (*Ensign*, Jan. 1999, p. 7).

At the end of a fireside where I had described the transforming, life-saving changes the Savior had brought into my life, I was asked if the change in my nature was instantaneous, or if it was a gradual process. My answer to both questions was yes. Yes, there was an instant change: a surprising peace, power, and confidence that came into my being from the moment I finally stopped trying to be my own savior and surrendered to Christ. But there was also a natural process of orderly growth that followed, as he gave me time and experience to make those changes in my character permanent.

It's my testimony that we can turn and walk away from bad habits, sins, even from addictions and compulsions, and, through the power of Christ, obtain victory from the very moment we put him first in our lives. I've experienced this myself, and I've seen it in the lives of others. It takes only a split second to commit our lives to God, but it requires time to live that commitment and make it a permanent part of our devotion. It takes time and patience to understand and fully apply his will, time to repent over and over as we stumble, time to conquer weaknesses and change habits. I like Stephen E. Robinson's explanation of what it meant when King Benjamin's converts testified "the Spirit of the Lord Omnipotent . . . has wrought a mighty change in us, or in our hearts, that *we have no more disposition to do evil*, but to do good continually" (Mosiah 5:2):

> From the moment of their conversion (or reconversion), the people of Benjamin changed their orientation and wanted righteousness rather than wickedness. It became their one goal. But that does not mean they achieved their goal instantaneously! It does not mean they never had another carnal thought or that they never subsequently lost any struggle against their carnal natures. At that moment, filled with the Spirit and clearly seeing the two paths before them, the people of Benjamin lost all desire to follow the path of evil. I feel the same way when I feel the Spirit, but I do not always feel the Spirit.
>
> That our disposition is good is proven by the fact that when we occasionally act otherwise, we feel bad about it, repent, and return to our pre-

vious heading toward righteousness *(Following Christ* [Salt Lake City: Deseret Book Co., 1995], pp. 41-42).

Each day as we face the struggles against our fallen natures, we may, from time to time, grow weary. Yet "we faint not" because even "though our outward [physical] man perish, yet the inward [spirit] man is renewed day by day" (2 Cor. 4:16). Our faith to persevere is sustained by the knowledge that Jesus Christ will "change our vile body, that it may be fashioned like unto his glorious body, according to the working whereby he is able even to subdue all things unto himself" (Philip. 3:21). In this day of fast food, ATMs, and almost instant everything, we must not only yield our entire selves to the Lord's wisdom, but trust his timetable as well. We can be assured that he will grant the changes and transformations we seek as quickly as he can prepare us because he has promised, "I will order all things for your good, as fast as ye are able to receive them" (D&C 111:11).

Nay, in all these things we are more than conquerors through him that loved us (Rom. 8:37).

EPILOGUE

"Wo unto him . . . that wasteth the days of his probation, for awful is his state" (2 Ne. 9:27).

We began this book with a little boy jumping in a mud puddle. It won't be many years before he discovers there is a much greater purpose to mortality than simply having fun. He, like each of us, will have to choose whether to rise above his fallen nature or to surrender to it, whether to master his mortality or allow his mortality to master him. What you do with your body today, tonight, and tomorrow is already determining what you'll be able to do with it during eternity.

A man was sentenced to ninety days in jail for disorderly conduct, a fairly minor offense. But instead of willingly paying his debt to society, or using those ninety days to study and accomplish something worthwhile, all he did was scheme and plot ways to escape. He finally devised a plan, and on the eighty-ninth day he escaped—for about five minutes. He was captured almost immediately, re-arrested, and sentenced to an additional eighteen months.

If we waste our mortality looking for ways to *escape* the imprisoning power of our fallen nature instead of finding ways to *rise above it*, to grow and become like Christ, we'll find ourselves serving an eternal sentence in a lower kingdom for all eternity.

There is an error in our comparison of the mortal body to a spacesuit. That is the fact that the astronaut will never become one with his suit, while we will. Eventually, through the resurrection, our spirit bodies will be merged with our physical bodies, or, as the scripture says, "inseparably connected" so that we may "receive a fulness of joy" (D&C 93:33). After the resurrection, we'll not only wear our physical bodies forever, but they'll become an essential part of *us* and we will become part of *them,* for "when separated, man cannot receive a fulness of joy" (D&C 93:34).

An immortal, resurrected body will be a free gift to every descendent of Adam, regardless of the kind of life that individual has lived (1 Cor. 15:22). But all resurrected bodies will not be the same in glory or capabilities. Paul explained that during mortality "all flesh is not the same flesh: but there is one kind of flesh of men, another flesh of beasts, another of fishes, and another of birds." He went on to explain that "so also is the resurrection of the dead" (1 Cor. 15:39, 42).

> There are also celestial bodies, and bodies terrestrial: but the glory of the celestial is one, and the glory of the terrestrial is another. There is one glory of the sun, and another glory of the moon, and another glory of the stars: for one star differeth from another star in glory (1 Cor. 15:40-41).

The scriptures teach that the *glory,* the *quality,* and the *capabilities* of the body we inherit for eternity will depend largely on the values and priorities we treasure during mortality and the way in which we treat our mortal bodies. Thus, in the resurrection, "ye shall receive your bodies, and *your glory shall be that glory by which your bodies are quickened.* Ye who are quickened by a *portion* of the celestial glory shall then receive of the same, even a *fulness.* And they who are quickened by a portion of the terrestrial glory shall then receive of the same, even a fulness. And also they who are quickened by a portion of the telestial glory shall then receive of the same, even a fulness" (D&C 88:28-31).

If we fail to train our bodies to live the celestial laws in *this* life, our resurrected bodies will be neither qualified nor prepared to dwell in a celestial glory in the *next* life. We'll be forever required to live in a lesser world, with fewer opportunities and joys, for "he who is not able to abide the law of a celestial kingdom cannot abide a celestial glory," just as "he who cannot abide the law of a terrestrial kingdom cannot abide a terrestrial glory" (D&C 88:22-23). (For an explanation of the differing capabilities and opportunities available to the different categories of resurrected beings, see D&C 76:50-112.)

When I shared my plans for this book with a devout Christian friend of another faith, she shared with me a prayer that she had uttered each morning for many years. I was deeply humbled by her desire to use her body in a way that would move her closer to his image and bring glory to her creator.

Bless my *eyes* this day to see the beauty in your creations and any opportunities where you need me to serve.

Bless my *ears* to listen to things both said and unsaid, that I might respond with love and compassion and hear needs even when unspoken.

Bless my *hands* to touch someone in a loving way and to perform some act of service that will encourage and strengthen.

Bless my *feet,* that they might stay on the straight path, moving forward toward thee.

Bless my *mouth* to give many smiles freely, and to say nothing that would wound or offend. To express praise and gratitude to thee and words of kindness that will lift and encourage another.

Bless my heart, that it might be soft and tender, pure in desire and concerned for others.

In February of 1996, while visiting in Utah, I had to drive from Orem to Ogden during the morning rush hour. On this day, traffic was crippled by a blinding snowstorm that slowed the movement on the congested freeway to less than five miles an hour. A journey that should have taken less than an hour and a half took almost four hours. During that trip, I passed dozens of cars that had been wrecked or had slid off the freeway into the ditch. I was very frightened by this experience—not so much because I was from Arizona and untrained for driving in snow, but because so many of those people with Utah licenses (who I felt *should* have known how to drive in snow) were colliding with each other and spinning off the freeway into the ditches.

As I thought about this experience later, I wondered if this might be what our lives look like to someone above, observing mankind making their way through the blinding blizzards of temptations and carnal nature that try to block our progress through mortality: wrecked and injured lives littering the straight and narrow path, millions of well-meaning people slipping into forbidden paths because of improper use of their bodies, stranded in the ditches of failure and poor choices.

I bear witness that it doesn't have to be this way. No Latter-day Saint has to be a victim of the fall or of carnal nature. We're *not helpless* against our fallen natures, nor against Satan's efforts. We're not *defenseless*. We know the gospel principles that will give us control of our bodies. We have access to the words of Christ, which will tell us all we need to know and do. We've been given the gift of the Holy Ghost to prompt and warn and guide us in every step we take to follow the Savior. And we have access to the power and grace of the Savior himself, which will *enable* us to do right when we are *willing* to do right.

During the past two years, our family has been involved twice in automobile accidents, with damage so severe that the insurance adjusters almost totaled our car. The cost of the repairs were so close to the value of the car that it would have been easier to scrap it than to repair it. Satan tries hard to make us feel like worthless junk when we make mistakes. I testify that Jesus never relegates anyone to the junk pile. No mistake or bad choice with your body can diminish your value in his sight or put you beyond his power to repair and make you like new—*better than new*.

Because of poor choices in using our bodies, many of us have slid off the straight and narrow path. We've slipped in the past, and we will likely slip again in the future. Some of us have been damaged or even landed in the ditch, helpless to get out by ourselves. Perhaps some of us have experienced these damages so often that we'd rather drive the rest of our lives in a battered car (or even remain in the ditch, battered and scarred) than to pray and confess another defeat. "This is a faithful saying, and is worthy of all acceptation, that Christ Jesus came into the world to save sinners" (1 Tim. 1:15). This is what he does, just as a doctor treats people who are ill or a mechanic repairs damaged cars. Don't be ashamed to let the Savior of mankind become your personal Savior. If you've been damaged by poor choices, but want to be made right again, he is both *willing* and *eager* to receive you and make you new. The Atonement and the power of Christ to change us into his image is infinite in its application, power, and possibilities, but it's worthless to you until it becomes personal. For you to be transformed into his image, you must change your perception to see Christ, not just as *our* Savior, but as *your personal* Savior. That's why the psalm says, "The Lord is *my* shepherd" (Ps. 23:1), and why Nephi testified, "I glory in *my* Jesus, for he hath redeemed *my* soul from hell" (2 Ne. 33:6).

Elder Jeffrey R. Holland spoke of this need for the Savior's atonement when he said:

> Do not be afraid of scars that may come in defending the truth or fighting for the right, but beware scars that spiritually disfigure, that come to you in activities you should not have undertaken, that befall you in places where you should not have gone. Beware the wounds of any battle in which you have been fighting on the wrong side. If some few of you are carrying such wounds . . . to you is extended the peace and renewal of repentance available through the atoning sacrifice of the Lord Jesus Christ. In such serious matters the path of repentance is not easily begun nor painlessly traveled. *But the Savior of the world will walk that essential journey with you.* He will strengthen you when you waver. He will be your light when it seems most dark. He will take your hand and be your hope when hope seems all you have left. His compassion and mercy, with all their cleansing and healing power, are freely given to all who truly wish complete forgiveness and will take the steps that lead to it" *(Ensign,* Nov. 1988, p. 78).

> Wherefore, whoso believeth in God might with surety hope for a better world, yea, even a place at the right hand of God, which hope cometh of faith, maketh an anchor to the souls of men, which would make them steadfast, always abounding in good works, being led to glorify God (Ether 12:4).

Making temples of our bodies will lead us not only to a better world in the *next* life, but to a better relationship between our bodies and our spirits and a better relationship with our Savior in *this* life. May you find here not the end of a book, but the beginning of a new life, an exciting new journey, a closer and more victorious walk with God, "which always causeth to triumph in Christ" (2 Cor. 2:14).

Thou wilt shew me the path of life: in thy presence is fulness of joy;
at thy right hand there are pleasures for evermore (Ps. 16:11).

INDEX OF STORIES AND ILLUSTRATIONS

INDEX

◆ H ◆

◆ S ◆

flesh, 139-42, 153; attacks our thoughts, 193-94; attacks through emotions, 210; at war with Christ, 139-40; become subject to his will, 153; cannot force us to listen, 58-59, 213; don't seek help, 240; filled world with filth, 194; his hold on hearts, 211; his hold over hearts, 211-13; need for Lord's help to resist, 181; our misery his goal, 142, 210, 221; persistent in temptations, 133; power over us, 142, 217; produces imitations, 251; puts thoughts in our mind, 193-94; ridicules weaknesses, 232, 262; says secret sins okay, 33; says faults are permanent, 173; says we cannot change, 239-40; seeks to blind us, 41-42, 210; seeks to enslave us, 140-42, 74; seeks to possess our bodies, 211-12, 325; slave to, 142; strategies, 221; use of memory of past, 173-77; use of pornography, 27-30; understands us, 213; whispers deceptions and temptations, 57-59, 224-25

SATISFIED, eyes of lust never, 27
SAVED, impossible without faith, 184
SAVIOR, should be personal, 263. See also Christ.
SCHOOL, mortal lessons, 88; pay attention to purpose of, 165-66; purpose of earth s., 139, 237. See also homework.
SCORCHED, by conscience, 190
SCOTT, Richard G., improper focus on past, 174, 176; slavery to Satan, 140-41, 212
SCRIPTURES, build trust, 39-40; duty to search and study, 181; compared to pacemakers, 222-26; effect on memory, 182-83; effect on mental polarity, 181-89, 196, 222-26;

effect on the mind, 181-89; essential to holiness and spirituality, 181; feasting on, 189; feelings toward s. = feelings toward Christ, 182; hearing alone insufficient, 64; hearing creates faith, 89-92, 184-86; help to know God, 183-84; importance of time spent with, 185; internalizing, 64, 186, 196-97, 222-25; keep us true, 187; love of s. a sign of discipleship, 186; prevent sinfulness, 223; provide nourishment, 186-89; reveal God's policies, 183; symbolized by iron rod, 44; word of Lord same as his will, 182; writing them in heart, 222-26
SEARCHING, with all heart, 232
SECOND BIRTH, see born again, change
SECRET, none from God, 91, 137, 191, 228; results of s. sins, 135, 212; revealed in judgment, 191, 228
SEE, see blindness and eyes
SEEDS, mental adultery plants, 27-30; must cultivate before harvesting, 184-85; planted by each choice and action, 116; planted by scriptures, 184-89; thoughts are s., 141-42, 163, 193
SEEK, with all the heart, 232
SELF-AWARENESS, through knowing God, 183
SELF-CENTEREDNESS, surrender to Christ-centeredness, 118
SELF-CONDEMNATION, 173-77. See also condemnation, guilt.
SELF-CONFIDENCE, see confidence
SELF-CONTROL, see control
SELF-DENIAL, 150-52. See also control.
SELF-DEPRECATION, don't substitute for repentance, 175
SELF-PUNISHMENT, don't substitute for repentance, 175
SELF-RELIANCE, more natural than